Nature in the Built Environment

Ambe J. Njoh

Nature in the Built Environment

Global Politico-Economic, Geo-Ecologic and Socio-Historical Perspectives

Springer

Ambe J. Njoh
School of Geosciences
University of South Florida
Tampa, FL, USA

ISBN 978-3-030-39758-6 ISBN 978-3-030-39759-3 (eBook)
https://doi.org/10.1007/978-3-030-39759-3

The cover photograph is of the Martyrs' Memorial Monument in Mekelle, Ethiopia.
Photographed by the author, Ambe Njoh, while serving as a United States Ambassadors' Distinguished Scholar, under the Ambassadors' Distinguished Scholars Program (ADSP) in the School of Architecture & Urban Planning at Mekelle University, Ethiopia (Fall 2019 – Spring 2020).

This Springer imprint is published by the registered company Springer Nature Switzerland AG
The registered company address is: Gewerbestrasse 11, 6330 Cham, Switzerland

This book is dedicated to the memory of my precious mother, Mammi Eni Diana Njoh, whose sojourn on Earth the Lord Almighty deemed necessary to halt on December 7, 2019 as I'd just scribbled the manuscript's concluding lines. Mother, while I sorely miss you, I take solace in the fact that you are savoring your eternal repose in God's tender loving arms.

Contents

Chapter 1
Environmental Stewardship and Built Space

Abstract There has been a rapid expansion of built space leading to unprecedented levels of urbanization throughout the world in the recent past. This has been matched by the rapid shrinkage and fragmentation of natural areas and resources. However, there are many gaps in knowledge of the exact nature of these diametrically opposed phenomena and how to go about resolving the problems they engender. This book contributes to efforts to close these gaps. Defining built space broadly to include all human settlements, the book interrogates efforts to promote nature in built space in different regions of the world. This chapter lays the foundation for, and provides a thumb sketch of, the discussion in the book.

1.1 Introduction

The discourse on the relationship between the natural environment and built space has intensified during the last decade. This discourse is no longer confined to academic quarters. Rather, it is becoming increasingly commonplace in the popular media, social forums, and among policy makers. The relationship is often characterized as conflictual—very likely a function of the erroneous tendency to view the natural and built environments as mutually exclusive. Ironically, this view is also prevalent in scholarly circles. The irony is evident in characterizations of the built environment as "a static creation, in a sense 'turned off' for a while from the circulation of nature" (Lamprecht 2016: 68). It can also be gleaned from studies seeking to determine whether people prefer the natural environment over built space or vise versa (e.g., Beute and de Kort 2018). Such studies view the two environments as mutually exclusive rather than complementary. Yet, their complementarity is indisputable. The complementary nature of both environments is glaringly obvious when one considers their indispensability for human survival. Humans need both the natural and built environments to survive. Elements of built space, including but not limited to the pieces of physical infrastructure that house people, commercial, religious, and health care activities lend credence to this assertion. Similarly, the soothing sounds of birds, the pleasant smell of flowers, and the sense of tranquility typically associated with the natural environment can significantly improve people's quality of life.

A. J. Njoh, *Nature in the Built Environment*,
https://doi.org/10.1007/978-3-030-39759-3_1

Advocating the harmonious co-existence of humans and nature is anything but novel. People have always shared space with the natural environment. Prior to the massive growth of human populations and the resultant urbanization trends, people were always surrounded by nature. Thus, the said coexistence is more natural than contemporary spatial organizational structures suggest. These structures are either products of modernist urban planning or seek to emulate principles of such planning. The structures have typically been oblivious to the need to protect nature. Consequently, efforts to develop built space have been cataclysmic by considering the destruction of natural features as unavoidable. Cases of such consideration are commonplace. Witness for instance, the common practice of indiscriminate tree-removal that typically precedes building and road construction projects. Also worth noting in this regard are efforts in so-called land reclamation initiatives that seek to transform swamps, lakes and other bodies of water into buildable land. The grave and disastrous environmental consequences of such initiatives are well established and need not be rehearsed here. Suffice to say that these initiatives are grossly incompatible with ongoing global efforts to promote sustainable development. This is because such development depends on the harmonious co-existence of the natural environment and built space. This dictates a need to develop built space in ways that seek to protect, not supplant or destroy, nature. This need is accentuated today, more than ever before, because of the rapid expansion of built space on the one hand and the speedy decline of the natural environment on the other.

The rationale for such co-existence is compelling; it is the need to harness the resources necessary for sustaining human life. As Fuller and Irvine (2010: 131) have observed,

> "the sheer rate and scale of human appropriation of natural resources has precipitated a biodiversity crisis currently being manifested in rapid rates of species extinctions, extensive transformation of the structure and function of ecosystems, and rapid alterations to the Earth's climate."

At the root of the current biodiversity crisis are human activities. Therefore, resolving the problem calls for human initiatives such as the aggressive re-introduction and preservation of nature in built space. In making an identical point, Fuller and Irvine (2010) contended that, people have the responsibility to ensure the growth and sustenance of nature in towns and cities because this is where most daily human-nature interaction occurs. The increasing expansion of built space has been matched by the rapid shrinkage and fragmentation of natural areas and resources. However, little is known on how to go about promoting sustainable economic growth, conserving biological diversity and maintaining ecological integrity in an increasingly urbanizing world (Mazzotti and Morgenstern 2014). To be sure, built space is not limited to urban areas; rather, it includes human settlements of all varieties.

The notion of nature co-existing with built space possesses both a logical and intuitive appeal. Trees and other plants add to, rather than subtract from, the quality of built space. They are sources of edible roots, fruits, and vegetables; in addition, they have direct health benefits as some contain medicinal ingredients. Also, they serve as avenues for recreational activities such as walking, jogging, and gardening.

Economically, surpluses from tree and plant products can be sold to generate income for communities, families and individuals. Also, public parks with well-maintained grass and trees serve as important avenues for socialization. Such facilities can provide an opportunity to preserve and conserve endangered species of plants and trees thereby serving a heritage function. Finally, nature in built space serves a crucial role in efforts to promote biodiversity, improve air quality and reduce CO_2 emissions.

The importance of trees and other natural elements in built space is magnified once we consider the rapid rate of urbanization that has been taking place throughout the world since the turn of the century. Presently, urban areas alone occupy about 2% of the planet's land area; and by 2030, they are projected to cover as much as 10% (Lamprecht 2016, para. 1; Vince 2014, 2015). In other words, by 2030 1.2 million square kilometers of additional biodiversity-rich landscapes would have been lost to urban construction alone (Lamprecht 2016, para. 1; Vince 2014, 2015). The accelerated urbanization trend observed since 2007 when, for the first time in human history, more than 50% of humanity lived in urban areas, shows no sign of slowing down. Most of the growth has resulted from rural to urban migration as opposed to human fertility. Thus, it is safe to focus meaningful efforts to ensure the survival of humans, other species and ecosystems in the Anthropocene on built areas. The main objective of such efforts should be to promote nature in built space. In other words, creating, maintaining and preserving natural resources in the built environment. However, there are vast gaps in knowledge on how to go about attaining this objective.

The main aim of this book is to contribute to efforts to close these gaps. It is premised on the belief that a good understanding of the status quo is necessary for the success of any effort to fill the gaps. Accordingly, the book scans the proximate and remote environments of global initiatives to create and protect nature in built space. Thus, the book conducts an environmental scan of the context of efforts to promote nature in built space in global perspectives. This necessitated the use of an analytical tool from the extant family of environmental scanning models (ESMs). The best-known of these is the "Strengths, Weaknesses, Opportunities, and Threats" (SWOT) matrix. This matrix, as its name implies, focuses on the strengths, weaknesses, opportunities and threats within the proximate and remote environments of an entity of interest. The matrix was initially designed for use in the business world by researchers at the Stanford Research Institute, California, USA in the 1960s (Humphrey 2005). Since then, it has been employed in a number of disparate fields. A few recent works, including Njoh (2017), Bas (2013) and Catron et al. (2013) have employed the matrix in the energy field. Yet, as an analytical tool, SWOT appears inadequate to the task of analyzing the political economy of nature in the built environment. In this regard, SWOT has been criticized on a number of grounds (Njoh 2017). Prominent in this regard has been the charge that SWOT lacks the versatility to handle multi-faceted contexts. Recognition of this shortcoming dictated the need to undertake a few modifications leading to what we herein label, PESTECH, which is an acronym encapsulating the political, economic, social, technological, ecological, cultural and historical context of an entity of interest. The modified model is employed in this book. The model shares many tenets with

STEEPLE, which is the acronym for the socio-cultural, technological, economic, ecological, political and ethical contexts of the entity being evaluated. As an analytical tool, it has been employed in evaluating the impact of a product on its users. PESTECH is a vastly improved variant of SWOT, STEEPLE and other existing environmental scanning models especially because of its sensitivity to context-specific factors such as ecology, culture and history. It is in order to examine each of the seven specific dimensions of the model—political, economic, social, technological, ecological, cultural, and historical—in turn.

Political factors include the political system and structure (e.g., totalitarian versus democratic), power structure, regional governments powers vis-à-vis the central administration, ratio of government to private ownership and control of business/utility companies, presence of state monopolies, government interference, level of influence from pressure groups on governments extent of political stability, level of government spending, safety, security, public protection, role/place of the military, state interference in the market, market regulations, trade agreements, tariffs or restrictions on imported commodities, taxes, clarity of procedures and laws governing imported goods' trade, institutional framework and governance structures, procedural formalities for obtaining permits for relevant activities, level of bureaucratic corruption, and government stability. An important subset of the political context comprises legal factors. These include all regulatory measures and laws that affect the functioning of the entity of concern. Basic questions that must be addressed when examining legal factors include but are limited to the following. To what extent are the rules, regulations, and laws implemented in a fair, just and equitable manner? Are the laws, rules and regulations standard throughout the country? Factors under the rubric of ethics include the ethical dimensions of the context in which the entity under consideration operates. The specific factors in question, must at a minimum, include morality, integrity, behavior, duties of citizens to themselves and others. What is considered 'good' or 'bad'? Economic factors include economic growth, employment policy and levels, inflation and interest rates, business climate, monetary policies, and consumer confidence, GDP, currency fluctuation, rate of inflation. Ecological factors include factors such as natural resource availability and accessibility, actions aimed at promoting sustainability, consumer appreciation, support of eco-friendly policies, fees and fines for natural resource use and exploitation as well as regulations affecting waste disposal. Technological factors include new inventions, development, rate of technology transfer, life cycle and speed of technological obsolescence, changes in information technology, and changes in mobile technology. The socio-cultural dimension includes factors such as indigenous cultural norms and values, income distribution/levels, demographic changes, labour, and social mobility, lifestyle changes, and educational levels. Historical factors are especially important because they constitute the foundation for all elements that affect any entity. This is particularly true when the entities of concern are polities. All polities, including countries and nation states have a history of some sort. This history, especially for formerly colonized states, have far-reaching implications for development policies and outcomes. Therefore, it can be exceedingly illuminating to consider this history in efforts to understand their current development profile. Appreciating a country's history is also necessary to understand its natural environmental policies especially as they relate to built space.

1.2 Main Questions

Three important questions are of centrality in this book. They include, for each major region, the following.

1. What is the institutional context of policies affecting nature in built space?
2. What are the implications of PESTECH factors for initiatives affecting nature in built space?
3. What specific steps have been taken to promote nature in built space?

These questions are tackled within the context of nature in built space as an element of commonly shared resources. Thus, it is necessary to summarize natural resources as a commons good.

1.3 Natural Resources and the Commons Question

Central to the concept of public goods is a seldom-acknowledged phenomenon, the 'self-governance of commons' (Ostrom 1990). Natural resources such as trees and other green areas, parks, and wetlands are by definition, public goods that require collective actions to manage. The 'collective actions' typically discussed in the literature are limited to those of members of the beneficiary communities. However, it is more instructive to broaden the scope of collective actions to encompass the actions of all stakeholders in the context of any commonly shared resource. These include the actions of suppliers, managers, regulatory agencies and consumers or users of the resource. Ostrom (1990) identified three problems likely to thwart such actions, namely the tragedy of the commons, the prisoner's dilemma, and the logic of collective action. The notion of 'tragedy of the commons' connotes the tendency of consumers or users of any resource to 'overexploit,' neglect the upkeep, and even behave opportunistically—that is, act as free-riders—of the resource (Hardin 1968). Critics of this concept, with Elinor Ostrom (e.g., 1990) at the forefront, have marshalled evidence to prove that this tendency is not as commonplace as previously believed. Instead, within many human communities, one is likely to find rules and institutions of the non-market variant that work to ensure the sustainable management and use of common pool resources. As a concept, the prisoner's dilemma holds that individuals are likely to seek to serve their own self-interest in any given situation (Poundstone 1993). Essentially, this assumes that individuals are wont to be suspicious of, and to avoid cooperating with, others. 'The logic of collective action,' which is the title of Mancur Olson's 1965 classic, essentially holds that concentrated minor interests will tend to be overrepresented while trumping diffuse majority interest because of the free-rider problem aforementioned. This problem tends to be intensified as groups grow larger. For Olson, individuals within any community summoned to undertake collective action towards achieving any given objective are likely to opt to 'free-ride.' This is contrary to Ostrom's position, which holds that members of common interest groups or communities, regardless of their size, are likely to act collectively to achieve their shared goals (Ostrom 1990). There are many reasons to believe that this holds true in the context of natural resources in built space.

One of these reasons is tied to the concept of stewardship. This concept has roots that are traceable to ancient civilizations, which assigned to humans the role of caretaker over natural resources. The concept is therefore of centrality in this book; the book acknowledges the fact that nature in built space involves multiple, and sometimes, disparate stakeholders. In this regard, the book deviates from conventional thinking in resource management and control. In contrast to this thinking, the book views the success of efforts to promote nature in built space as a function of multiple factors, including but not limited to, the political, economic, social, technological, ecologiclal, cultural and historical contexts. Analyzing these contexts in global perspectives would prove to be invariably revelatory. It would reveal that all societies recognize the importance of nature and built space harmoniously co-existing. Also, there is a universal recognition that public interests trumps private preference when it comes to nature in built space. As Wright and Boorse (2017) have noted, even in Western societies where individual land ownership is the norm, there is a recognition of the fact that 'ownership' is essentially a temporary phenomenon as any given piece of land is always guaranteed to outlive its 'owner.'

Despite the universal nature of certain practices and preferences, many politico-economic, socio-cultural and geo-ecological factors are unique to certain regions. There are vast differences in the ways in which these factors impact efforts to promote nature in built space by region. However, there are many gaps in knowledge of these differences. How, for instance, does the dominance of an aspect of culture such as Islam affect efforts to promote nature in built space in Middle-Eastern countries? How does the communitarian ethos, an element of African indigenous culture affect such efforts in sub-Saharan Africa? How is such an ethos received in the context of promoting and protecting nature in built space in the avowed capitalist societies of North America and Western Europe? This book is intended to contribute to efforts to these and cognate questions. It accomplishes this task by analyzing the impact of political, economic, social, technological, ecological, cultural and historical factors on efforts to promote nature in built space. The units of analysis are the major United Nations Environmental Programme (UNEP) regions of the world. The focus is on four specific substantive areas in environmental policy, namely forestry, water, food, and energy. Although these substantive areas or resources are important everywhere, their degree of importance differs by region contingent, especially, but not exclusively, upon ecology, geography and availability.

1.4 Forests in Built Space

Forestry in built space falls under the general rubric of afforestation. Broadly defined, this includes conventional forestry initiatives and 'greening activities' such as tree planting and gardening or horticulture in urban areas. Urban greening programs are increasingly being implemented as a strategy to promote environmentally sustainable urban milieus. The activities that go into efforts in this connection are elaborate and often require the participation of many entities including individuals, members of the beneficiary communities, groups, government agencies, non-governmental organizations, and private commercial organizations. Thus, afforestation, like many other areas of environmental management, has many stakeholders. These entities invariably interact with each other; and their effectiveness, according to inter-organizational relations theory, depends on the quality and magnitude of this interaction. This interaction is of centrality in the book. The aim is to investigate the impact of the interaction and other proximate factors on efforts to promote nature in built space.

1.5 Water in Built Space

The second substantive domain of interest in the book is water. The impact of PESTECH factors on water resource management cannot be overstated. By its very nature as an indispensable resource, water tends to involve the greatest number of stakeholders; its quality and quantity is also vastly affected by geo-ecological factors. Water transcends familiar geo-political boundaries, including conventional borders circumscribing communities, regions and even countries. This explains the proliferating number of studies dedicated to analyzing water in terms of its quantity, quality and distribution networks. These studies fall into two categories, namely potable water supply and agricultural water control. I have personally completed many studies belonging to the first category, including a number of community-based water supply projects in Cameroon (see e.g., Njoh 2002, 2006). I have also played a leading role in similar projects elsewhere such as Kisumu, Kenya (see, Ananga et al. 2017). Examples of works in the second category include, Dungumaro and Madulu (2003). This work uncovered a number of justifications for community participation (CP) in water systems management. In this purely qualitative study, the authors accentuated the importance and the indispensability of CP in improving water resource management. In another study, Apipalakul and colleagues (2015) investigated the role of CP in dealing with problems facing communities in the Pong River Basin in Thailand. Like communities in other river basins, those around the Pong River Basin commonly encounter problems arising from their joint activities such as agriculture, irrigation and industry in the basin area. The authors' articulation of the CP concept suggests that it is more than simply a framework for understanding collective purposeful processes. Rather, it is also a useful strategy for conflict resolution in water basin communities. In yet another study on the role of CP in water resource management, Boakye and Akpor (2012) summoned evidence to demonstrate CP's versatility. However, they were quick to acknowledge the fact that meaningful CP remains a challenge especially in historically impoverished communities. The avowed purpose of their study was to determine the extent to which residents of communities in South Africa consider their participation meaningful in a water catchment forum. Yet, it is important to note that CP in water resource management is not confined to developing countries. Rather, it is increasingly becoming a universal strategy for ensuring the sustainable management of water resources throughout the world. In the European Union, this trend is exemplified by legislative actions such as the European Water Framework Directive; and in the United States, it is manifested by the U.S. Clean Water Act (Carr et al. 2012). Each of these pieces of legislation mandates public and stakeholder participation in water resource management. Geo-ecological factors are also well-established determinants of water quality and quantity. In this regard, the arid regions, such as the Middle East, tend to suffer the most from problems of water stress. Yet, these regions do not have any monopoly over water stress problems For instance, the European Environment Agency (EEA) recently noted that a third of EU-member countries are facing low water availability problems (EEA 2019). Countries in this category boast less than 5000 cubic meters of water per head per year. The problem is particularly grave in Southern Europe, where a country like Malta, with only 100 cubic meters of water per head per year is located.

1.6 Energy Generation and Supply

The third substantive domain of interest in the book is energy. Energy is a natural product whose production and use are conditioned by politico-economic, ideological and socio-cultural as well as historical factors. Thus, the PESTECH is apropos as an analytical framework in the context of energy as an element of nature in built space. In concert with the book's concern with sustainability, the analysis will focus more intensely on renewable energy. There is a resurgence of interest in renewable energy that is increasingly attracting the participation of members of the general public. The level of interest is heightened in both the developed and developing world. In the former, the concern is typically with renewable energy as a more environmentally-friendly alternative to conventional energy. In the latter, interest in renewable energy is provoked mainly by the need for energy cost minimization given the resource scarcity problems of developing countries. Over the years, residents of these countries have developed creative strategies for making energy more affordable. Prominent among these strategies is community participation (CP). It is employed mainly as a strategy to pool the resources necessary to defray the high cost of energy supply equipment. In India for instance, this strategy permitted villages such as Sagar, Mousuni and Sundarbans in East Bengals, to develop their own off-grid power systems (Dwivedi, Online). However, it must be noted that these systems involved not only the input of the villagers alone but also that of the Indian Government, which furnished half of the cost in each case. Another example of such an initiative comes from Gambia. Here, residents of Batokundu Village marshalled resources such as money, labour, land and enlisted the technical support of a German NGO, to complete a windmill electrification project for the village (Hathaway 2010). Two more examples of projects exemplifying this trend come from Kenya. The projects are located in the Mount Kenya area and specifically in the villages of Kathama and Thima (Hathaway 2010). With the technical support of the British NGO, Practical Action, the two communities joined forces to tap power from a micro-hydro plant to meet their electricity needs. While the NGO helped the communities to navigate the cumbersome process of securing government permits, members of the communities contributed labor, building materials, and land required for the project.

1.7 Food Security Issues

The last substantive area of interest in the book is food. Despite improvements occasioned by modern technology, much physical space is still required to produce food. Accordingly, human population increases cause a lot of ecological footprint because of the huge demands these increases make on land (Deelstra and Girardet 2000). The concept of ecological footprint is vital in illustrating the complex ways in which human settlements, particularly cities, affect the natural environment. Conventional footprint analysis typically assumes that activities in these settlements depend on the supply of a finite quantity of vital resources such as land, food and water. The quantity of these resources required to sustain the population of a city is that population's ecological footprint on the earth.

A more obvious and ominous problem has to do with the scarcity of livable space. This problem is especially acute in densely populated urban areas. This accentuates the need to maximize the use of available space. The U.N. Food and Agricultural Organization (FAO) has been at the forefront of efforts in this regard. In one of its Factsheets on "urban and peri-urban horticulture," the organization describes a strategy "to boost the overall supply of horticultural produce to the world's developing cities" (FAO 2019, para. 1). The strategy calls for low-income households to supplement their food supply by developing micro-gardens. Micro-gardens permit urban residents to produce their own vegetables, roots, tubers and other food in very little space. These gardens have three main attractive features, namely ability to maximize the utility of extant spaces, mobility, and environmental friendliness. The space maximization prowess of micro-gardens is evident in the fact that they require very small and typically un-used spaces such as balconies, patios and rooftops that already exist in/or around buildings. A micro-garden can operate on an area as small as one square meter. The mobility aspect of micro-gardening is a function of the fact that crops are grown in movable containers. Thus, instead of planting directly into the ground, the crops are planted in arable soil and manure in containers such as plastic-lined wooden crates, custom-built tables, and used tires. The environmental-friendly attribute of micro-gardening relates to the fact that it depends on rainwater and household waste to function. In this case, rainwater is harvested in containers from where it is collected and used for watering the plants.

The FAO has provided funds to assist the state and municipal authorities in poor developing countries to launch micro-gardens. These gardens have helped to meet the nutrition needs, and serve as a viable income source, for low-income families. Efforts in this connection have registered enormous success. For instance, an FAO-supported program designed to promote micro-gardens in Caracas, Venezuela helped 10,000 poor, barrios residents to grow their own leafy vegetables, cabbages, pumpkin, tomatoes, and eggplant (FAO 2019, para. 4). Similar FAO-supported programs have proved equally successful in some African countries, notably among which are Gabon, Namibia, Niger, Senegal and Rwanda. Empirical data suggest that micro-gardens serve not only as a source of food, but also a source of income for low-income urban households. For instance, a study on Senegal showed that only 35% of the micro-garden produce is consumed by the producers while the rest is sold (FAO 2019: para. 4).

1.8 The Concept of Ecological Footprint

The importance of ecological footprint is amplified by the increasingly domineering stature of cities today. In fact, 2000 marked the first time in human existence that more than 50% of humanity lived in urban centers. One upshot of this is the fact that cities are consuming an increasingly voluminous quantity of natural resources on the one hand, and generate much waste, on the other. Both phenomena result in destroying the habitat of several species. A more obvious consequence of the growth and proliferation of cities is the multiplicity of the number of mouths that must be fed. This magnifies the need for efforts to promote the development and preservation of natural resources, particularly those that can serve as sources of edibles in human settlements. Such efforts are anything but novel. Often undertaken under the rubric of urban agriculture, these efforts have contributed to supplementing the food inventory in human settlements for a long time. The following examples are illustrative (Deelstra and Girardet 2000). In the 1980s and 1990s, urban agriculture was the source of 30%, and 40%, respectively of the dollar value of agricultural production in the United States. A few urban centers boast levels of urban agricultural activities that render them self-sufficient with respect to some food items. For instance, Singapore, which is entirely built-up, has always produced enough meat to feed its population, meanwhile Bamako, the capital of semi-arid Mali, has been able to meet its vegetable need without resorting to importation (Deelstra and Girardet 2000).

Of interest in this book is the impact of socio-cultural factors in facilitating the functioning of agricultural activities in built space, particularly urban centers. Socio-cultural factors have always been instrumental in food security initiatives. This is true in contemporary as in ancient civilizations. For instance, communities throughout ancient Africa operated highly sophisticated and effective systems of community food programmes. The best-known testament to this assertion is provided in the Holy Bible (see e.g., Genesis 41). Here, one finds a vivid description of the elaborate granaries that were created by Joseph—known as Joseph's Granaries—to store grain in seasons of abundant harvest to be consumed during seasons of famine in Egypt. In Zimbabwe, ancient Shona communities are on record for operating community food programmes under the rubric of 'Zunde Ramambo,' a Shona appellation for the Chief's Grain Reserves (Swikepi 2011). The programme entailed households within a given community jointly farming a designated parcel of land to meet their future collective food needs. Food from such jointly operated farmland was usually stored in silos at the chief's palace and only distributed in time of low harvests or to families in need such as widows, widowers, orphans or the physically challenged. With the introduction of the capitalist mode of production and the commensurate growth and proliferation of urban centers, the importance of communal food programmes waned. However, programmes bearing a passing resemblance to these were instituted from the mid-1980s to the early-1990s to address the adverse effects of the World Bank/International Monetary Fund-initiated Structural Adjustment Programmes (SAPs) (Sseguya et al. 2013). Among other things, SAPs prescribed trade liberalization, privatization and especially the elimination of food subsidy programmes. Thus, the purpose of community food programmes has remained the same—to meet people's food need—from time immemorial. This need has certainly not been confined to Africa although it took until the 1970s for international authorities to acknowledge food shortage as a global problem. Upon this acknowledgment, the World Food Conference of 1974 underscored the right-to-food as a human right (Adams et al. 2007). Also emerging from this renewed consciousness has been the determination to maintain an acceptable level of security. The notion of food security has been described as a situation in which people within any given community have easy, safe and sustainable access to food (Adams et al. 2007). Efforts to guarantee food security in the United States have included *inter alia*, the operation of community food banks. In doing so, American authorities have tapped on an important aspect of the country's Judeo-Christian culture, namely the requirement to 'love thy neighbor as thyself.' Food banks exist throughout the country. The book analyzes how culture and other PESTECH factors affect efforts to create and preserve natural resources such as food in built space.

1.9 Book Outline

The book contains eight chapters including this introductory chapter. Chapter 2 discusses environmental stewardship and its role in efforts to promote nature in built space. Six chapters—Chaps. 3 through 8—follow this, and focus respectively on the major regions of the world approximately corresponding with the regional grouping scheme preferred by the United Nations Environmental Programme (UNEP). As employed in this book, the regions include, Sub-Saharan Africa, Middle-East and North Africa, West Asia, Asia and the Pacific, North America, and Latin America and the Caribbean. Each of the chapters begins with an introduction that includes a brief description of the region's main environmental problems affecting built space in terms of their magnitude and intensity. This is followed by a discussion of culture, beliefs, and other factors affecting efforts to promote nature in built space in each region. The focus is particularly on belief systems, especially religions and their influence on the desire and willingness to develop and maintain natural resources in built space. Following this in each chapter is a discussion of efforts in each of the four substantive domains, forest, energy, water and food in each of the major target regions.

References

Adams AE, Swisher ME, Monaghan KN (2007) Defining community food security. Document #AEC 383, Series of the Agricultural Education and Communication Department, IFAS Extension, University of Florida.

Ananga EO, Njoh AJ, Ananga GO, Pappas C (2017) Examining the relationship between community participation and water handling hygiene practices in the informal neighborhoods of Kisumu, Kenya. Habitat International (in press)

Apiplakul C, Wirojangud W, Ngang TK (2015) Development of community participation on water resource conflict management. Procedia Soc Behav Sci 186:325–330

Bas E (2013) The integrated framework for analysis of electrical supply chain using an integrated SWOT-fuzzy TOPSIS methodology combined with AHP: the case of Turkey. Electr Power Energy Syst 44:897–907

Beute F, de Kort YAW (2018) Thinking of nature: associations with natural versus urban environments and their relations to preference. J Landsc Res 44(4):374–392

Boakye MK, Akpor OB (2012) Community participation in water resources management in South Africa. Int J Environ Sci Dev 3(6):511–516

Carr G, Bloschl G, Loucks DP (2012) Evaluating participation in water resource management: a review. Water Resour Res 48(11). https://doi.org/10.1029/2011WRO11662.

Catron J, Stainback GA, Dwivedi P, Lhoka J (2013) Bioenergy development in Kentucky: a SWOT-ANP analysis. Forest Policy Econ 28:38–43

Deelstra T, Girardet H (2000) Urban agriculture and sustainable cities. In: Bakker N, Dubbeling M, Gundel S, Sabel-Koshella U, de Zeeuw H (eds) Growing cities, growing food: urban agriculture on the policy agenda. ZEL, Feldafing, pp 43–66

Dungumaro EW, Madulu NF (2003) Public participation in integrated water resources management: the case of Tanzania. Phys Chem Earth 28:1009–1014

EEA (2019) Renewable energy in Europe: key for climate objectives but air pollution needs attention. European Environmental Agency. Briefing No. 13/2019

FAO (2019) With micro-gardens, urban poor "grow their own." urban and peri-urban horticulture. Factsheet 6. Retrieved on 30 July 2019 from: http://www.fao.org/ag/agp/greenercities/pdf/FS/UPH-FS-6.pdf

Fuller RA, Irvine KN (2010) Interactions between people and nature in urban environments. In: Gaston KJ (ed) Urban ecology. Cambridge University Press, Cambridge, pp 134–171. https://doi.org/10.1017/CBO9780511778483.008

Hardin G (1968) The tragedy of the commons. Science 162(3859):1243–1248

Hathaway T (2010) What is driving dams in Africa. World Rivers Rev 25(4):5–5. (December 2010)

Humphrey A (2005) SWOT analysis for management consulting. SRI Alumni Newsletter (SRI International) 1:2005

Lamprecht M (2016) The role of the built environment in human life. Eur Spat Res Policy 23(2):65–78

Mazzotti FJ, Morgenstern C (2014) A scientific framework for managing urban natural areas. Document No. SSWEC74, Wildlife Ecology and Conservation, UF/IFAS Extension, University of Florida IFAS Extension. Accessed 17 Apr 2017 at: http://edis.ifas.ufl.edu/uw112

Njoh AJ (2002) Barriers to community participation in development planning: lessons from the Mutengene (Cameroon) self-help water project. Commun Dev J 37(3):233–248

Njoh AJ (2006) Determinants of success in community self-help projects: the case of the Kumbo (Cameroon) water supply project. Int Dev Plan Rev 28(3):381–406

Njoh AJ (2017) The SWOT Model's utility in evaluating energy technology: illustrative application of a modified version to assess the sawdust Cookstove in sub-Saharan Africa. Renewable an Sustainable Energy Reviews 69:313–323

Ostrom E (1990) Governing the commons: the evolution of institutions for collective action. Cambridge University Press, Cambridge

Poundstone W (1993) Prisoner's Dilemma. Random House, New York

Sseguya H, Mazur RE, Njuki JM, Owusu FY (2013) Determinants of participation and leadership in food security groups in Southeast Uganda: implications for development programs and policies. J Rural Commun Dev 8(1):77–97

Swikepi C (2011) Community participation and food security in rural Zimbabwe: the case of Marange area in Mutare District. Unpublished Master's thesis, University of Fort Hare

Vince G (2014) Adventures in the Anthropocene: a journey to the heart of the planet we made. Chatto & Windus, London

Vince G (2015) Whom is nature for? Am Sch (November 15, 2019)

Wright RT, Boorse DF (2017) Environmental science: toward a sustainable future. Pearson, New York

Chapter 2
PESTECH and Nature in Built Space: Analytical Framework

Abstract Of centrality in this book is the issue of nature in built space. As a subset of the broader conversation on the natural environment, and a matter of scholarly and professional interest, this issue has a major flaw. It lacks a well-developed analytical framework. Consequently, meaningfully discussing the issue has often proved difficult at best. Cognizant of this, this chapter proposes a framework that promises to facilitate efforts to analyze matters in the environmental policy field. The framework is fashioned after the environmental scanning models (ESMs) that have been employed to evaluate the environment of business organizations. In particular, the framework focuses on the political, economic, social, technological, ecological, cultural and historical (PESTECH) factors of the proximate and remote environments of efforts to promote nature in built space. This chapter paints a vivid picture of the framework, which guides the book's entire discussion.

2.1 Introduction

The discourse on nature in built space is a subset of the broader conversation on the natural environment writ large. A common flaw in this conversation is the absence of well-developed analytical frameworks. The natural environment as well as the principles and rules that govern its treatment are often believed to be universal. This is erroneous because people's perception and relationship with the natural environment are conditioned by many factors. Foremost among these are the political, economic, social, technological, ecological, cultural and historical (PESTECH) factors of the proximate environment of the object being analyzed. Considering these and related factors is akin to scanning the environment of the object under examination. PESTECH is fashioned after PESTLE, a popular environmental scanning model (ESM) used in the business world. It is typically employed to analyze the political, economic, social, technological, environmental and legal environment of business organizations (Business 2016). PESTLE also assumes other identical acronyms such as PESTEL and PEST. It is often employed in tandem with cognate tools such as SWOT (Strengths, Weaknesses, Opportunities, and Threats). The model has proved versatile in analyzing the macro-environmental factors with real or potential implications for the performance of business organizations.

A. J. Njoh, *Nature in the Built Environment*,
https://doi.org/10.1007/978-3-030-39759-3_2

Paradoxically, the rich insights of the model and other variants thereof, have seldom been employed to analyze the politico-economic, socio-cultural and geo-ecological context of the natural environment. Yet, such insights are necessary to uncover real and potential facilitators and impediments to environmental policy-making. Cognizant of this, this book employs the PESTECH framework to analyze initiatives to promote nature in built space in major regions of the world. An original coinage, PESTECH is a member of the environmental scanning family of models. The main objective of the chapter is three-fold. The first is to discuss environmental scanning models (ESMs) and how they have been employed to analyze environmental phenomena in the relevant literature. The second is to discuss PESTECH as the analytical framework employed throughout the book. The final objective is to discuss an important but oft-ignored feature of the culture component of ESMs, namely belief system, as a leading determinant of people's view of, and relationship with the natural environment.

2.2 ESMs in Studies of the Natural Environment

As mentioned above, the best-known environmental scanning models ESMs such as SWOT, PESTLE and PEST were originally developed for the business world. Consequently, they have had to be modified to suit contexts outside this world. However, only a few studies have undertaken the necessary modification of the models for use in studies of the natural environment. The few efforts in this connection have focused mainly on one aspect of the natural environment, namely energy. My work, which employs SWOT to determine the suitability of the sawdust cookstove in sub-Saharan Africa is among the most recent of these studies (see Njoh 2017). Another example is the study by Cole and colleagues (2005) that employs the SWOT matrix to examine energy storage and transmission levels. Yet another example is the study by Rutz and Janssen (2007), which also employs SWOT to analyze renewable energies in Munchen, Germany. As an ESM, SWOT has also been used in a natural environment-related area to assess the performance of energy corporations. One such study focused on the US-based General Electric Corporation (GE) (GE, Online). It employed the model to analyze the macro-environment of the organization. A few of the studies employing the environmental scanning analytical framework have had countries as their units of analysis. Such studies lend credence to the regional orientation adopted in this book. Exemplifying studies in this group is the work of Bahn and colleagues (2013), which employs the SWOT model to analyze Canada's energy and climate policies. With a nation, namely Canada, as the empirical referent, this study sought to determine the extent to which sectoral policies, particularly those dealing with energy and climate, align with federal and provincial/territorial development priorities. By reformulating a conventional environmental scanning model to focus on a whole nation as opposed to an organization, Bahn and colleagues (2013) were able to highlight implications for energy sector performance of the country's politico-administrative structures and stakeholder interests. The SWOT matrix permitted the researchers to identify certain weaknesses characteristic of the country's energy sector. These include inequalities in the distribution of energy resources and the conflictual nature of the country's energy and climate policies. The case of energy policies that encourage fossil fuel production illustrates this conflictual nature. Here, it is necessary to note that fossil fuel production conflicts sharply with climate policies intended to reduce greenhouse gases.

The afore-cited study by Bahn and colleagues (2013), and others like it (e.g. Markovska et al. 2009) deviate considerably from conventional environmental scanning studies that focus on specific organizations. Yet, it is necessary to note that the focus—national as opposed to organizational—is only one of the many differences between the two types of studies. Another, and perhaps more important distinction, is the purpose for which environmental scanning is undertaken in the two situations. In the business world, environmental scanning is typically designed to help business leaders react better to forces in the relevant environment of their organizations. Thus, environmental scanning in the business world is designed to serve reactionary objectives. In contrast, environmental scanning targeting larger units of analysis such as countries and regions is designed to accomplish proactive planning purposes. In practice this typically calls for activities such as scanning to: a) collect data on a phenomenon of interest; b) promote knowledge of that phenomenon; and c) take steps to maximize the utility of its positive attributes while minimizing the impact of its negative features. Seen from this perspective, environmental scanning models can best be characterized as planning tools.

Environmental scanning models have also been employed in non-business settings to analyze macro-natural-environmental policies, especially at the regional level. The works by Terrados and colleagues (2007), and Catron and colleagues (2013) are illustrative. The former analyzed the energy portion of the strategic plan of Jaén, an administrative province of southern Spain. The natural resource of primary interest was renewable energy, particularly solar and biomass. In employing an ESM model, the researchers argued that such a model is ideal for use in redesigning regional energy systems. This is because ESMs have proven their versatility in guiding territorial strategic processes. Furthermore, these models are designed to be sensitive not only to the current but also future state of affairs.

The re-tooling of ESMs from the business sector for use in analyzing natural environmental phenomena has assumed many formats. One of these simply adds one or more dimensions to capture extant realities pertaining to a specific phenomenon of the natural environment. The work of Catron and colleagues (2013) that modified the conventional SWOT matrix to include Analytic Network Processes is illustrative of this trend. With the modification, the matrix became SWOT-ANP—that is, Strength, Weaknesses, Opportunities and Threats-Analytic Network Processes—and was used to elicit valuable information to implement a bio-mass energy project in Kentucky, USA. Thus, with the modification, an otherwise conventional business evaluative tool proved exceedingly effective in informing concerned authorities on, whether or not, to proceed with the implementation of a renewable energy project. In fact, the ESM-based assessment revealed a lack of support from the potential project beneficiaries. They believed that bioenergy production in Kentucky had more negative than positive implications. The analysis also unveiled information pointing to the inextricably intertwined nature of factors affecting the bioenergy production process.

Another noteworthy modification to the conventional business sector ESMs for use in analyzing phenomena in the natural environment has to do with adjusting certain nomenclatures. One example of this is the use of public policy outcomes as a proxy for business sector products. Two notable works, one by Chen and colleagues (2013), and another by Xingang and colleagues (2013) exemplify this trend. The study by Chen and Colleagues focused on three countries, Japan, South Korea and Taiwan. By employing an ESM, particularly SWOT, the researchers were able to gain a deeper appreciation of the strengths, weaknesses, opportunities, and threats of energy policies of three countries. Their main revelation is that the countries share the common goal of combatting the problem of deficiencies in their domestic fossil fuel supply. In particular, the three countries have initiated vigorous policies to develop renewable energy. The more notable contribution of this work to the extant body of knowledge on ESMs is its broadening of the scope of analysis to include multiple countries. Note that the SWOT model, as used in the business world, has typically focused on one organization at a time. Efforts to adopt the model in the context of natural environmental policymaking, as noted earlier, have also been on a country-by-country basis. Broadening the scope of the model to permit the simultaneous analysis of more than one country at a time has many advantages. One of these is that researchers are able to make recommendations with implications for a region as opposed to a single country. For example, broadening the scope as Chen and colleagues did in the afore-cited study, they were able to recommend informed policy actions to the three countries they studied. In particular, they recommended for the countries to enter cooperative initiatives that enabled them to establish themselves as exporters of renewable energy. This recommendation has implications for countries in similar situations elsewhere.

The study by Xingang and colleagues (2013) also re-tooled the conventional ESM to employ policy in lieu of a product as called for in the business world. The specific use to which the modified model was put is itself creative. It was employed to achieve two objectives, namely to assess the external and internal development environment of China's shale gas; and to evaluate the shale gas development status of China based on the four dimensions of SWOT, namely strength, weaknesses, opportunities and threats. In addition, the model provided a framework for building strategies to enhance the development of the Chinese shale gas industry. Above all, it permitted assessment of the comparative advantages of shale gas vis-à-vis alternative energy sources. Foremost among the advantages of shale gas is its low carbon emissions capability.

Another innovative aspect of research employing modified ESMs to analyze natural environmental phenomena is to evaluate a technology's capabilities. This constitutes a marked deviation from the original focus of ESMs, namely business organizations. However, studies of this genre remain rare. Two of the few studies with this focus are respectively by Cole and colleagues (online) and Rutz and

Janssen (2007). Cole and colleagues employed a business ESM to assess various energy storage technologies. They were interested in the strengths, weaknesses, opportunities and threats (SWOT) of competing technologies, including two traditional types, namely batteries and flywheels as well as three relatively new ones, viz., Superconducting Magnetic Energy Storage (SMES), supercapacitors and hydrogen. An important revelation of their study is that choice of a technology depends on the parameters of the intended application. More importantly, the choice must be conditioned by the cost, time and place of intended application. Some technologies may show enormous promise for the future but are of little value in the present. For instance, while hydrogen shows promise for the future, it is currently inefficient and prohibitively costly.

In their study, Rutz and Janssen (2007) used the standard, Strengths, Weaknesses, Opportunities, and Threats matrix to assess the sustainability of biofuels in comparison to other energy sources. The study revealed biofuels to be ideal especially in the face of the rapidly decreasing global supply of fossil energy resources. The use of an environmental scanning model permitted the researchers to understand energy, particularly biofuels from a more comprehensive perspective. For instance, the analysis produced evidence suggesting that biofuels are not without blemish. More specifically, it was revealed that while biofuels have many positive attributes, they are also saddled with a number of problems. On the positive side, biofuels can contribute to reducing gas emissions and guarantee a steady supply of energy. On the negative side, their production entails an extravagant use of land and financial resources. In an effort to promote understanding of these crucial attributes of biofuels, Rutz and Janssen (2007) conducted a strength, weakness, opportunities and threats (SWOT) analysis of the technology. In particular, they analyzed biofuels initially, in comparison to fossil fuels and then, in relation to each other. A noteworthy point about this study is the fact that it employed an ESM to analyze a whole region, namely the European Union. Thus, the use of such a model to analyze United Nations Environmental Programme (UNEP) regions as done in this book is not unreasonable. The next section characterizes PESTECH, the specific ESM employed in the book in terms of its composition and scope.

2.3 PESTECH, the Conceptual Framework

The acronym, PESTECH, as employed here, stands for the political, economic, social, technological, ecological, cultural and historical contexts of any entity under evaluation. It is an environmental scanning model (ESM) or tool for analyzing organizations, policies, technologies and cognate entities. The model and other models in this family have been employed to gauge the impact of a product on its users. In the business world, such models permit executives to anticipate future trends for their businesses within the broader context of its macro- and micro-environments. It is a vastly improved variant of conventional models that are very familiar to business organizational analysts. While the conventional ESMs typically examine four or less dimensions, PESTECH evaluates as many as seven dimensions. It is in order to examine each of these dimensions—political, economic, social, technological, ecological, cultural and historical contexts—in turn.

2.3.1 Political Factors

Political factors include the political system and structure (e.g., totalitarian versus democratic), power structure, and regional government powers vis-à-vis those of the central administration. Also lumped under the rubric of political factors are, ratio of government to private ownership and control of business/utility companies, presence of state monopolies, government interference, level of influence from pressure groups on governments, and extent of political stability. The following are also considered elements of political power: level of government spending, safety, security, public protection, role/place of the military, state interference in the market, market regulations, trade agreements, tariffs or restrictions on imported commodities, taxes, clarity of procedures and laws governing imported goods' trade, institutional framework and governance structures. In addition, no meaningful discussion of political factors in the context of ESMs would exclude such elements as procedural formalities for obtaining permits for relevant activities, level of bureaucratic corruption, and government stability. An important subset of political factors comprises legal elements such as regulatory measures and laws that affect organizational functioning. Basic questions that must be addressed when examining legal factors include, but are limited to, the following. To what extent are the rules, regulations, and laws implemented in a fair, just and equitable manner? Are the laws, rules and regulations standard throughout the country? Factors under the rubric of ethics include the ethical dimensions of the context in which the entity under consideration operates. The specific factors in question, must at a minimum, include morality, integrity, behavior, duties of citizens to themselves and others. What is considered 'good' or 'bad'?

It is not enough to simply understand the role of PESTECH's political components. To be helpful in the context of the analytical framework for this book, it is necessary to appreciate the component's essence for efforts to promote nature in built space. The elements with the most direct impact on these efforts include the following (also see Table 2.1):

- Urban planning laws and regulations;
- Government involvement in community development;
- Size of government budget;
- Level of government subsidies;
- Environmental policies;
- Government stability.

Table 2.1 PESTECH factors and their implications for nature in built space

Item	PESTECH component	PESTECH elements relevant to nature in built space	Sample implications for nature in urban space
1.	Political	Government stability/instability; Corruption level; Urban planning laws and regulations; Government regulation and deregulation; Government involvement in community development; Environmental policies Recycling standards Level of government subsidies; Intergovernmental relationships; Import-export regulation/restrictions; Size of government budgets.	Government stability is very important for nature in built space. Instability, especially when it involves wars often results in destroying urban infrastructure, including green infrastructure. Good governance, including the absence of corruption is necessary to develop and maintain nature in built space. Urban planning laws and regulations are necessary to designate and protect space for nature in built areas.
2.	Economic	Size; Growth rate; Availability of credit to individuals, local governments and communities; Level of disposable income; Propensity of people to spend; Gross domestic product (GDP) trend; Level of employment.	Developing and maintaining nature in built space requires considerable financial resources. The availability of such resources depends on the climate prevalent in a country or any polity for that matter.
3.	Social	Population size and growth rate; Education level; Minorities; Crime levels; Birth rates; Death rates; Number of marriages; Number of divorces; Immigration and emigration rates; Life expectancy rates; Age distribution; Wealth distribution; Social classes; Per capita income; Family size and Structure; Lifestyles; Health consciousness; Average disposable income; Attitude towards government; Attitude towards work; Attitudes towards customer service; Attitudes towards emigrants.	Social factors such as those listed here impact the extent to which nature in built space can be used. The factors also have implications for the sustainability of green infrastructure. Members of the active population are necessary not only to use, but also develop and maintain green infrastructure in built space. People's attitude towards health and well-being impact how much of the green facilities such as exercise trails, parks and swimming are utilized.

(continued)

Table 2.1 (continued)

Item	PESTECH component	PESTECH elements relevant to nature in built space	Sample implications for nature in urban space
4.	Technological	Technology incentives Automation R&D activity Technological change Access to new technology Level of innovation Technological awareness Internet infrastructure Communication infrastructure Life cycle of technology	Technological innovation has been very impactful on nature in built space in recent times. For instance, the ability to use very little space such as roof tops and porches to grow large quantities of food is enhanced by technological innovation.
5.	Ecological	Weather Climate Climate change Pressures from NGO's Natural disasters Air and water pollution	Ecological factors are determinants of the specific type of nature that can be developed and maintained in built space in any region. Some specie of trees cannot survive in some regions.
6.	Cultural	Cultural norms and values; Sex roles and distribution; Religion and beliefs; Attitudes towards green products; Support for renewable energy; Racial equality; Attitudes towards retirement; Attitudes towards leisure time; Ethical concerns.	Culture is exceedingly important for the survival of nature in built space. One particular element of culture, namely religion, has demonstrated its ability to shape people's thinking about the natural environment. In particular, religion is an important determinant of environmental stewardship.
7.	Historical	History of the culture of the region or country; History of major formal institutions; History of concern with the issue at hand—in this case, the natural environment in built space; History of organized government—in former colonized states, this has to do with the state's colonial history.	The history of a region or country is a strong determinant of its present and future state. It also constitutes a viable predictor of its present and future actions. Hence the need to understand the environmental history of any entity.

Urban Planning and Environmental Policies Urban planning also goes under the following appellations: city planning, urban and regional planning, town planning, rural planning, and urban development. It is a *bona fide* aspect of the political component of PESTECH for many reasons. The most important of these is the fact that it is more than a technical process, which employs scientific principles to design and develop human settlements or built space writ large. It is, arguably above all, a political process with the objective of reconciling disparate societal interests in the use of land and the location of land use activities in built space. To accomplish these objectives requires the enactment and implementation of a series of planning laws and regulations. These tools, especially zoning regulations, dictate the timing, extent and location of land use activities, including urban forests, trees, farms, and recreational facilities. To the extent that this is true, planning laws and regulations must be considered the political element of PESTECH with the most consequential impact on efforts to promote nature in built space.

Government Involvement, Size of Government Budget and Level of Government Subsidies It is impossible to overstate the importance of government involvement as an element of the political component of PESTECH in efforts to promote nature in built space. Such support can be quantified in terms of the extent of financial commitment by the government to local green infrastructure projects. Yet, this is not the only measure of government influence on such projects. Other meaningful measures are the size of the government's budget and the level of subsidies provided by the government to support such projects. The resources necessary to realize such initiatives, especially the technical expertise and money, are often beyond the means of local communities. Governments in developed countries are fully aware of this problem and have taken the necessary steps to address it. As an example, the Federal Government in the United States, through its Environmental Protection Agency (EPA) operates funding programs that help municipal authorities throughout the country to realize local green infrastructure projects. In rationalizing its involvement in such initiatives, the EPA (Online, para. 1), drew attention to the fact that "lack of funding is consistently cited as a barrier to the implementation of green infrastructure." This is consistent with the EPA's long-standing policy of supporting initiatives that promote sustainable and resilient communities (Box 2.1).

In the case of green infrastructure, the EPA (2014) makes a persuasive case for one specific form of green infrastructure, namely 'green storm water infrastructure.' The thrust of the case is that investments in such green infrastructure, particularly green roofs, roadside rain gardens, and increased tree canopy—collectively grouped under low impact development—possess the potential to support the mission of a range of Federal programs.

> For example, roadside rain gardens and trees can improve walkability while also reducing stormwater run-off from streets. In turn such green amenities can also help attract private investments that revitalize corridors and neighborhoods. Additionally, strategic greening of land or rooftops in a community can also be instrumental in managing flooding, providing opportunities for growing food locally and improving energy efficiency (EPA 2014: para. 1).

Consequently, despite the will to do so, local communities are unable to develop such infrastructure. This dictates the need for support from the state or government, including municipal, state and federal governments.

Government Stability/Instability This is also a crucial element of the political component of PESTECH that has far-reaching implications for efforts to promote nature in built space. Government instability, especially when it involves wars, often result in destroying urban amenities, including green infrastructure. Also consequential for the sustenance of nature in built space is Good governance, which includes the absence of corruption. If nothing else, good governance ensures, for instance, that resources earmarked for green infrastructure are not misappropriated.

Box 2.1: Federal Government Support for Local Green Infrastructure Projects: The Case of the US EPA

The United States' Federal Government initiative known as "The Green Infrastructure Collaborative" constitutes an acknowledgment of the fact that, on their own, local communities are incapable of bearing the total cost of green infrastructure projects. The many agencies involved in this collaborative initiative share the belief that local green infrastructure projects portend benefits that transcend local administrative and geographic boundaries. In fact, the benefits reverberate nationally and internationally. This is especially true on account of the fact that green infrastructure constitutes a viable tool for building resilience to climate change impacts, including but not limited to increased heavy rainfall and heat island effect. The agencies comprising the Partnership for Sustainable Communities include the U.S. Environmental Protection Agency (EPA), U.S. Department of Transportation (DOT), and U.S. Department of Housing and Urban Development (HUD) – along with U.S. Department of Agriculture (USDA), U.S. Department of Interior (DOI), U.S. Department of Defense (DOD) and U.S. Department of Energy (DOE). Each of these offer unique expertise and possess the resources necessary to assist communities plan for, design and implement local green infrastructure projects. (Source: EPA 2014).

2.3.2 Economic Component

This includes the size of the economy as well as other indicators of economic prosperity such as economic growth, employment rate, inflation and interest rates, business climate, monetary policies, and consumer confidence, GDP, currency fluctuation, rate of inflation. The economic component also speaks to questions such as the availability of credit to individuals as well as communities and local governments, level of disposable income and the propensity of individuals to spend. These economic measures influence the ability of local governments and communities to invest in nature in built space. They also affect both the willingness and ability of people to use green infrastructure, particularly parks, urban forests, beaches, rivers and lakes. Where the economy is characterized by downward trends, it is unlikely that governments and local communities would possess the resources to develop these facilities. As stated above, developing green infrastructure requires large amounts of financial and other resources. Similarly, without a prosperous economy, people find it difficult to meet the cost of enjoying the use of green space and its commensurate facilities.

2.3.3 *Social Component*

The social context for the purpose of the present discussion includes safety issues, the demographic characteristics, population mix and size, the trends of this population, the growth rate, the income distribution, and lifestyles of members of the population in the proximate area of any built space under consideration. The social context also includes a concern for matters such as labour, and social mobility, lifestyle changes, and literacy or educational levels. More importantly, it encompasses demographic attributes such as age, gender, and marital status. These factors influence the type of green infrastructure that needs to be developed, and the extent to which this infrastructure is utilized. In this regard, families with young children are more likely than those with older ones to use public parks. Also, it is reasonable to expect married couples to make greater use of parks and cognate recreational facilities than single individuals, particularly those without children.

2.3.4 *Technological Factors*

These include new inventions, development, rate of technology transfer, life cycle and speed of technological obsolescence, changes in information technology, and changes in mobile technology. Of prominence in the context of the present discussion is, hi-tech agriculture, particularly the use of advanced technological techniques that accelerate plant growth and significantly increase food yields. As Tyagi (2018) observed, hi-tech agriculture has significantly increased food productivity through revolutionized farming. Technological developments are already having serious implications for initiatives to promote nature in built space. Some of the most impactful of these have been in the area of urban farming or agriculture. Here, advanced technology is currently permitting farming techniques such as hydroponics and aeroponics. These techniques have made it possible to use very little space to grow large quantities of food. In particular, hydroponics and aeroponics make it possible to grow vegetable in water and air respectively. This nullifies the use of soil. New technology has also made it possible to grow temperate vegetables in tropical areas; it has also produced disease resistant plants.

2.3.5 *Ecological Components*

These include elements such as natural resource availability and accessibility, actions aimed at promoting sustainability, consumer appreciation, support of eco-friendly policies, fees and fines for natural resource use and exploitation, regulation affecting waste disposal. More importantly, the ecological component includes natural features of an area or region, including the vegetation, rivers, lakes, topography, rainfall, precipitation, climate, freshwater and groundwater availability. It also includes five important environmental factors: exposure to hazardous substances in the air water, soil, and food; frequency of natural disasters, climate change and occupational hazards. These factors influence the composition of green infrastructure in any locale. For example, the factors affect the type of trees that can be grown in any country or region.

2.3.6 Cultural Component

The cultural context incorporates norms, customs, values, belief systems, and lifestyles. These affect many aspects of built space, including the choice of residential location vis-à-vis other land use activities. Culture also influences people's relationship with nature. In particular, it influences people's relationship with animals. For example, the cow is a sacred animal in Hindu culture. Hence, the fact that cows are not an uncommon sight in Hindu-dominated societies such as India. Also, traditionally cattle-herding cultures have always exhibited an animal-friendly disposition. The common sight of animals such as horses, cows, donkeys and camels on streets in countries such as Ethiopia, Niger, Burkina Faso is illustrative. A more important dimension of culture—in fact, the most important for the purpose of this book—is religion. As discussed in the remainder of this chapter, religion constitutes a crucial determinant of environmental stewardship.

2.3.7 Historical Component

Despite its obvious importance, the historical component is often omitted in environmental scanning models (ESMs). Yet, history has never ceased to constitute a defining characteristic of any entity, including but not limited to organizations, countries and regions. This history is more often than not, a reliable predictor of the future of these entities. This is especially true in the case of formerly colonized states, whose contemporary development profiles are almost invariably linked to their colonial past. Therefore, it can be exceedingly illuminating to consider this history in efforts to understand their environmental policy initiatives, particularly with respect to nature in built space. Here, I hasten to note that many towns and cities in erstwhile colonies developed around nucleuses created by colonial authorities. More importantly, the laws and regulations that guide the production of built space in these countries are relics of their colonial past.

2.4 Beliefs, Ideology and Environmental Stewardship

People's beliefs and ideological persuasions are instrumental in shaping their thinking on, and treatment of, the natural environment. In short, few factors surpass people's beliefs as shaped by religion, in determining how they relate to the natural environment. Eminent ecological ethicist, John Grim (1997) provides some support to this assertion when he draws on the authoritative pronouncements of environmental philosopher J. Baird Callicott to advance the following argument. Religious traditions play a significant role in answering ecological and cosmological questions about the manner in which humans treat the Earth's living systems. However, as Grim is quick to concede, "the roles of many of the world's religions in understanding human relations with ecosystems have not been extensively developed" (p. 139).

Also worth noting for the purpose of the discussion in this book is the fact that people's view of community participation and its role in natural resource management is a function of their beliefs and ideological proclivities. These, in turn, are usually rooted in, or derived from, religion, which may be of the organized or unorganized variant. Therefore, accounting for religious beliefs and practices is one way by which the discourse on the natural environment in global perspectives can be significantly advanced.

It is in this light that the present chapter undertakes to dig deeper into the relationship between religion and environmental stewardship. In particular, it highlights the various ways in which different religions and their respective belief systems affect people's perception of the natural environment. Initially environmental stewardship as a function of religion, ideology, and beliefs are weaved into a single model to demonstrate their interconnectivity.

The natural environment as well as the principles and rules governing its treatment are conditioned by intangible phenomena such as religion, belief systems, ideology, norms and culture. They are also a function of tangible phenomena such as geographical features. Thus, appreciating different religious doctrines, belief systems, cultures and traditions is a logical starting point for understanding why people in different regions of the world, and sometimes within the same region or country view and relate to the environment differently. Figure 2.1 incorporates the variables articulated here in a single structural model. It posits beliefs, ideology, norms, culture and tradition as predictor variables upon which the propensity for positive interpersonal relationships and environmental stewardship depends. Also, it posits interpersonal relationships as a determinant of environmental stewardship. This relationship will become obvious with some appreciation of the world's major religions, belief systems as well as their respective doctrines and principles.

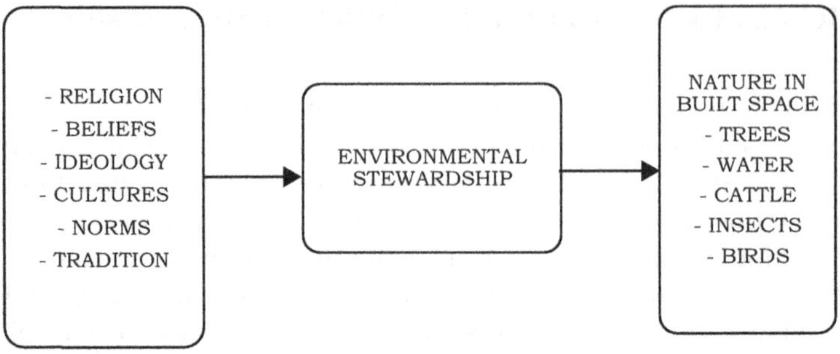

Fig. 2.1 Model of environmental stewardship as functions of religion, beliefs, ideology; and as predictors of nature in built space

2.4.1 Religious Doctrines and Environmental Behavior

Religion is arguable the most prominent aspect of culture. This is especially true when religion is taken to mean a system of complex and multifaceted beliefs, rituals and moral teachings. There is a multitude of such systems throughout the world. Table 2.2 summarizes vital information on the world's multiple religions. To the extent that people can be taken at their word, religiosity is pervasive throughout the world today. According to a recent British Broadcasting Corporation's (BBC) online article, "eight out of 10 people around the world consider themselves religious" (BBC 2019). As high as this proportion appears, it is indicative of a decline and does not include people who are affiliated with unorganized religions. The implications of this for the way people relate to the natural environment are far-reaching. This is especially because religious beliefs are well-known for permeating the core fabric of human existence. Essentially, people's thoughts and behavior are influenced in one way or another by their religion, especially what they believe about themselves in relation to other human beings and creatures in the universe. As Lynn White stated in his widely-cited, if only controversial article, "The Historical Roots of Our Ecologic Crisis," people's treatment of the natural environment depends on what they believe about themselves vis-à-vis the things around them (White 1967). Thus, some appreciation of the major religions of the world, including their fundamental beliefs can prove exceedingly useful for efforts to understand people's relations to the natural environment. How people view themselves as human beings vis-à-vis other creatures can be placed on a continuum. On one extreme of this continuum is the view of humans as the dominant creature with a preordained right to control and subdue all other creatures; on the other extreme is the view of humans on an equal footing with other creatures as co-inhabitants of the Earth. The view of humans as the dominant creature is commonplace; it appears to draw its inspiration from Judeo-Christian doctrine. It can also be blamed for fueling resistance to programs designed to promote environmental stewardship. However, Christian theologians find this charge abhorrent and have been quick to advance arguments designed to ferociously challenge it. In one such argument, Bergstrom (2014) underscores the

Table 2.2 World religions by number of adherents and view of humans vis-à-vis the natural environment

Item	Religion	Number of adherents	Share of world's population	Sub-groups and geographic location	Concept of humans vis-à-vis nature
1	Christianity	2,173,180,000	31%	The most widely spread religion. Present in all Western countries, sub-Saharan Africa, Latin America, and Caribbean. Catholics (50%), Protestants (37%), orthodox (12%).	God created humans to have dominion over all other creatures and to exploit/use these for their own pleasure.
2	Muslim	1,598,510,000	23%	Sunnis (87–90%), Shia (10–13%). Predominant in the Middle East and North Africa.	Humans are the deputy of God on earth. They are required to maintain the earth in accordance with God's will.
3	No affiliation	1,126,500,000	16%	Atheist (20% in the US), no particular faith given.	Humans are just one specie among tens of millions on this planet. Therefore, humans are not superior to the other species with which they co-habit the earth.
4	Hinduism	1,033,080,000	15%	Almost all (94%) live in India	Humans are just one of the elements of the universe which the Supreme God created for the benefit of all. God is present in everything including nature. So humans are obligated to treat it with utmost care.
5	Buddhism	487,540,000	7%	Half (50%) live in China.	Emphasizes the avoidance of killing, preaches compassion towards all living things, advocates reverence for trees and other life forms. In addition, Buddhists meditate in natural surroundings. Therefore, within the framework of Buddhist thinking, humans must protect the environment because they need it to survive.

(continued)

Table 2.2 (continued)

Item	Religion	Number of adherents	Share of world's population	Sub-groups and geographic location	Concept of humans vis-à-vis nature
6	Indigenous religions	405,120,000	6%	These are mainly located in Africa and other non-Western areas.	Humans are just one of the myriad species that populate the universe. They are a piece of nature and have not created, and cannot create, any part of it. Therefore, they have no right to destroy any part of it.
7	Judaism	13,850,000	0.2%	Mainly located in the US (41%) and Israel (41%).	Humans were created to be superior to all other creatures of the earth. God endowed them with powers that were not given other creatures: high mental ability that permits them to think and reach conclusions on their own; God also gave them the power of speech. They were created to be superior to, and given dominion over, other creatures.
8	Other religions	58,110,000	1%	Include Baha'I faith, Taiosm, Jainism, Shintoism, Siknism, Tenrikyo, Wicca, Zaroastrianism, etc.	Some of these religions, e.g. Taiosm have no omnipotent being beyond the cosmos. So, for them, there is no being that created and controls the universe. The natural environment is not something for humans to master and dominate but their partner in an everlasting relationship.

Source: Culled from worldometer.info (Online), Adherents.com (Online)

need to read Biblical passages on man's position as the dominant creature within the context of the Bible in its entirety. Doing so, allows us as humans to see ourselves as creatures sharing a common planet with other creatures; as other creatures, humans depend on the Earth for life-support.

Thus, rationality, and in fact, our survival instinct makes environmental stewardship mandatory. Yet, there is a plethora of evidence suggesting that Christians consistently exhibit less environmental stewardship than non-Christians, especially followers of indigenous religions. Therefore, people's values and beliefs regarding nature can explain their view of environmental stewardship. The basis of these beliefs and values can be found in religious doctrine, which differ from one religion to the next. A number of empirical studies have since marshaled evidence to support White's hypothesis that religious doctrines and beliefs affect people's environmental behavior (see e.g., Sherkat and Ellison 2007; Eckbergy and Blocker 1989).

2.5 Conclusion

The discourse on environmentalism has intensified and grown more universal in the recent past. However, many gaps remain; the most prominent of these being the paradoxical lack of a well-developed analytical frameworks. Typically, it is assumed that the natural environment and the principles and rules that govern its treatment are not only well-established but universal. Yet, nothing could be further from the truth. People's perception and relationship with the natural environment is a function of several, sometimes disparate factors. The most important of these, according this book, are the political, economic, social, technological, ecological and cultural (PESTECH) characteristics of the proximate environment or the proximate environment of the object being analyzed. This chapter has weaved these components to constitute a framework for analyzing the issue of nature in built space. This is tantamount to scanning the environment of initiatives to promote nature in built space in different regions of the world. Such an analytical exercise is certainly not novel; analytical frameworks of the sort developed and employed here are commonplace in the business world. These frameworks belong to the family of analytical tools known as environmental scanning models (ESMs). What is novel is the use of the framework in the discourse of nature in built space writ large. An important contribution of this chapter is its development of a robust framework for analyzing efforts to promote nature in built space in global perspectives. It is hoped that the framework will prove versatile for efforts to analyze environmental initiatives elsewhere. Such frameworks have already proved versatile in efforts to understand the macro-environmental factors with real or potential implications for business organizational performance.

References

BBC (2019) Religion can make us more environmentally friendly—or not. Retrieved, March 15, 2019 from: http://www.bbc.com/earth/story/20170206-religion-can-make-us-more-environmentally-friendly-or-not

Bergstrom JC (2014) Subdue the Earth? What the Bible says about the environment. An Online Paper for the Apologetics Resource Center. Retrieve, April 4, 2019 from: https://arcapologetics.org/culture/subdue-earth-bible-says-environment/

Business (2016) Scanning the environment: PESTEL analysis. https://www.business-to-you.com/scanning-the-environment-pestel-analysis/. Accessed 25 Oct 2019.

Cole S, Van Hertem D, Meeus L, Belmans R (2005) SWOT analysis of utility-side energy storage technologies. Proceedings of the 5th WSEAS/IASME International Conference on Electric Machines, Tenerife, Spain, Dec. 16–18 (pp. 446–451). https://www.researchgate.net/publication/237327418_SWOT_analysis_of_utility-side_energy_storage_technologies. Accessed 18 Feb 2020

Cole S, Van Hertem D, Meeus L, Belmans R, Energy storage production and transmission level: a SWOT analysis. https://scholar.google.com/scholar?hl=en&q=Energy+storage+production+and+transmission+level%3A+a+SWOT+analysis+&btnG=&as_sdt=1%2C10&as_sdtp. Online Accessed 3 Oct 2015

Eckbergy DL, Blocker TJ (1989) Varieties of religious involvement and environmental concerns.: testing the Lynn White thesis. J Sci Study Relig 28(4):509–617

EPA (2014) Federal agency support for green infrastructure collaborative. Article by the US Environmental Protection Agency (EPA). https://www.epa.gov/sites/production/files/2015-10/documents/federal-support-for-green-infrastructure-collaborative_508.pdf. Available Online. Accessed 15 Nov 2019

EPA (Online) Green infrastructure funding opportunities. Informational online posting by the Environmental Protection Agency. https://www.epa.gov/green-infrastructure/green-infrastructure-funding-opportunities. Accessed 15 Nov 2019

Grim JA (1997) Indigenous traditions and ecological ethics in "Earth's insights." Special theme issue on J. Baird Callicott's "Earth's insights". WORLD 1(2):139–149

Markovska N, Taseska V, Pop-Jordan J (2009) SWOT analysis of the National Energy Sector for sustainable development. Energy 34:752–756

Njoh AJ (2017) The SWOT Model's utility in evaluating energy technology: illustrative application of a modified version to assess the sawdust cookstove's sustainability in sub-Saharan Africa. Renew Sustain Energy Rev 69:313–323

Rutz D, Janssen R (2007) Biofuel SWOT-analysis. WIP Renewable Energies, Munich

Sherkat DE, Ellison CG (2007) Structuing the religion-environment connection: identifying influences on activism. J Sci Study Relig 46(1):71–85

Tyagi S (2018) Hi-Tech Agriculture a solution for food security March 2018 conference: International Conference on Research and Extension for Sustainable Rural Development at: Rural Development Academy (RDA) Bogra, Bangladesh

White L (1967) The historical roots of our ecologic crisis. Science, New Series 155(3767):1203–1207

Chapter 3
Nature in Built Space in Sub-Saharan Africa

Abstract There are many political, economic, social, technological, ecological, cultural and historical (PESTECH) factors affecting efforts to promote nature in built space in sub-Saharan Africa (SSA). Because of its checkered history, especially as it concerns colonialism, the region's policies are heavily influenced by historical factors. This is particularly the case with policies relating to nature in built space, which are a subset of urban or town planning policies bequeathed to indigenous authorities by departing colonial masters. This chapter highlights the traces of colonialism not only in its treatment of the historical factors but also in its analysis of the political, economic and technological contexts of efforts to promote nature in built space. In doing so, the chapter demonstrates the intensity of colonial influence on urban and peri-urban forestry, and farming, water supply and energy in the built environment in SSA.

3.1 Introduction

The focus of this chapter is sub-Saharan Africa, the region extending from south of the Sahara Desert to the southernmost part of the African continent. It also includes the islands of Cape Verde, Equatorial Guinea and Sao Tome and Principe in the Atlantic Ocean; and Madagascar, Seychelles and Réunion in the Indian Ocean (See Fig. 3.1). The region is punctuated by sporadic highlands that stretch from Cameroon, the home of Mount Cameroon, through Rwanda and Burundi, through Zimbabwe to Tanzania, the locale of Mount Kilimanjaro. It comprises arid and forest zones with an average rainfall of about 100 mm per annum. The rainfall in the forest zone, including the Congo Basin and the West African coast, is significantly higher at 4000 mm per year; this zone is generally moist and humid. The East, South-East and South-West regions of the continent are characterized by tropical savannah land of grass and sparsely distributed trees. A common thread running through these countries is their European colonial heritage. European colonial authorities are credited with crafting the nucleus around which built space, especially urban areas, in the region have developed. This point is worth noting because of its implications for policy and other initiatives to develop and manage nature in the built environment in the region. Sub-Saharan Africa has the dubious reputation of being the least developed major region in the world. However, this often leads to

A. J. Njoh, *Nature in the Built Environment*,
https://doi.org/10.1007/978-3-030-39759-3_3

Fig. 3.1 Map of Africa showing sub-Saharan Africa in dark green

the misleading view of the sub-continent as a place frozen in history with little more than wildlife to offer. As promoted in travel brochures and magazines, the region is endowed with an abundance of natural resources, including rivers, lakes, and lush forests, where an assortment of wildlife roams uninterruptedly. But travel magazines have no monopoly in portraying this exotic, if only derisory, image of the region, which deliberately ignores its built features. Just some four or so decades ago, an otherwise reputable American newsmagazine had reported that Africa was devoid of any history of built space (Njoh 2007). The magazine, said notable architectural historian, Labelle Prussin (1974), was relaying the message from its photographers who had been dispatched to capture images of African architecture for a picture story on world architectural styles. The photographers had returned empty-handed, after a sojourn in Africa to report that they could not find anything of architectural value to photograph. "All we could find," the photographers reported, "were a bunch of mud huts" sparsely punctuating vast expanse of forested land or grass fields (Prussin 1974: 183). Was this portrayal accurate–but it is certainly not–there would be no need to identify and analyze factors affecting nature in built space in sub-Saharan Africa.

Archaeological evidence suggests that sub-Saharan Africans were involved in city building initiatives long before the Greeks and Romans (Njoh 2006). In fact, at the time Prussin was writing, the region boasted several large towns and cities, of which some, such as Lagos and Kinshasa, already boasted populations in excess of one million. Today, the region is home to many large cities as well as megacities, that are facing the same need and obstacles to develop and maintain green space as their international peers. However, it is necessary to understand that while superficially, issues relating to greening built space may be alike, a closer inspection reveals that their nature and magnitude are contextually determined. This chapter seeks to shed light on the unique identity of, and factors that influence or can potentially influence, nature in built space in sub-Saharan Africa (SSA).

The chapter contains nine sections; Sect. 3.2 follows this introduction, and presents some background information on the political, ecological/environmental, social/demographic, economic, technological and historical contexts of the SSA region. Sect. 3.3 discusses major strands of African indigenous traditional beliefs and practices and shows how these impact environmental stewardship in general, and nature in the built environment in particular. Sect. 3.4 discusses the land question in SSA and its impact on efforts to develop and manage nature in built space. Sect. 3.5 discusses water resources and the many strategies that have been summoned to ensure its availability in built space. Sect. 3.6 is concerned with afforestation and greening initiatives in urban areas throughout the region; while Sect. 3.7 examines efforts to guarantee food security through different forms of urban farming. Sect. 3.8 identifies and analyzes different strategies for improving energy access in built space; while Sect. 3.9 concludes the chapter.

3.2 Background

Contemporary sub-Saharan Africa (SSA) includes the following countries (in alphabetical order): Angola, Benin, Botswana, Burkina Faso, Burundi, Cabo Verde, Cameroon, Central African Republic, Chad, Comoros, Congo (PR), Congo (DR), Cote d'Ivoire, Djibouti, Equatorial Guinea, Eritrea, Ethiopia, Gabon, Gambia, Ghana, Guinea, Guinea-Bissau, Kenya, Lesotho, Liberia, Madasgascar, Malawi, Mali, Mauritania, Mauritius, Mozambique, Namibia, Niger, Nigeria, Reunion, Rwanda, Sao Tome and Principe, Senegal, Seychelles, Sierra Leone, Somalia, South Africa, Sudan, Swaziland, Tanzania, Togo, Uganda, Western Sahara, Zambia, Zimbabwe. The region, which includes all 50 countries south of the Sahara Desert, is a product of a triple heritage incorporating elements drawn from indigenous African cultures, Islamic doctrine and Western/Judeo-Christian traditions. All of these elements affect the way people in the region perceive the natural environment in general and nature in the built environment in particular.' The exotic accounts of early European explorers and Christian missionaries portrayed Africa as the 'dark continent.' Yet, Africa was anything but a dark continent even prior to the arrival of Arabs on the continent more than a millennium before the early visits by European traders and explorers in the fourteenth century. Africans had a sophisticated sense of nature and had developed principles governing people's relationship to it. Prominent in this connection are principles, albeit unwritten, that implored Africans to treat nature as God's gifts to people and all other earthly creatures. These principles are at the core of African traditional religion. This, perhaps because of its unorganized nature, has been pejoratively characterized by visitors to the continent from time immemorial.

The pejorative characterization of indigenous African tradition has always been the preamble for efforts to supplant it with imported alternatives. In the case of the Arabs who had settled in Mombasa, Zanzibar and other areas on the coasts of East Africa, their aim was to supplant indigenous African religion and traditional practices with Islamic equivalents (Njoh and Akiwumi 2012; Mazrui 2004). With respect to nature, one principle, the relationship of people to land, is noteworthy. Here, there were irreconcilable differences. While Islam viewed land as a commodity that can be owned and sold, African indigenous religion and culture recognized the sanctity of land as a special creation of God that can neither be owned nor sold by anyone (Njoh and Akiwumi 2012).

Arguably the most conspicuous and indelible marks on built space in the sub-Saharan African region are a legacy of its colonial experience. Therefore, colonialism, the vehicle through which Western values were initially introduced in the region, deserves more than passing attention in any meaningful analysis of nature in this space. Although Europeans had been present in Africa since the fourteenth century, they did not begin amassing territory on the continent until the Berlin Conference of 1884/85. Subsequent to this conference, which officially launched the colonial era in Africa, European countries embarked on a frantic scramble to acquire territories on the continent. Barely a decade-and-a-half following the conference, most of the African continent was already under colonial rule. By 1912, nearly all of the continent had been acquired by European colonial powers (History World Online). Germany, the first to acquire a territory on the continent, controlled five colonies, including Cameroon, Togo, Namibia, and Tanganyika (present-day, Tanzania). These territories were, however, lost to allied countries as part of the outcome of World War I. With sixteen and eighteen colonies, respectively, Britain and France controlled the lion's share of the continent. For about a century colonial powers transplanted politico-administrative principles and practices as well as spatial and environmental planning models from their respective home countries to the colonies. In addition, they employed different strategies to implement colonial policies. Two of these, direct rule, and indirect rule strategies, typically associated with French and British colonial enterprises, respectively, are among the most discussed in the colonial/post-colonial literature. These strategies have enormous implications for citizen participation (CP) in environmental and other policy making milieus. However, irrespective of their nationalities, all European colonial powers in Africa worked fervently to promote Christianity. In fact, some analysts have opined that Christianity was employed as a tool to facilitate the colonial enterprise (Njoh 2006). More accurately, though, the two operated reciprocally; colonialism relied on Christianity while Christianity's success hinged tightly on the fruitfulness of the colonial enterprise. As Richard Gray (1982: 60) observed, "Christianity. .. made its rapid advances precisely because its emissaries, the missionaries, were so closely linked with the whole apparatus of colonial rule."

Political Context Most countries in the sub-Saharan African countries gained independence in the 1960s. Since then, the countries have, almost without exception, faced many nagging problems with the following seven being the most dominant (Lepapa 2017): poor governance, corruption, unemployment, population growth, insecurity, droughts, and famine. Many of the countries, such as Cameroon, Equatorial Guinea, Mozambique, the Central African Republic, and Angola, to name just a few, have been led by strongmen or authoritarian regimes under single-party rule. In addition, the countries have been facing many political problems. Prominent in this regard is the problem of weak and inept institutions that lack policy implementation capabilities. This problem is compounded in the francophone countries, because they have religiously maintained the highly centralized governance structures from their erstwhile French colonizers. In addition, there is the problem of political instability coupled with armed internal conflicts. These conflicts have resulted in the destruction of not only public infrastructure but also green spaces and natural resources in built space. More importantly, the conflicts have been incriminated for causing extensive biodiversity loss in both the built and natural environments. The severe damage visited upon environmental resources in both the natural and built environments by armed conflicts in the twentieth and twenty-first centuries are illustrative. Consider the impact of the following specific armed conflicts and their effect on natural resources. The civil wars in the Congo Democratic Republic from 1998 to 2000, in Rwanda in 1994, in Somalia in 1991, and the war between Ethiopia and Eritrea from 1961 to 1993, and from 1997 to 2000 reported by Enzler (2006) are illustrative. In Cameroon, government forces have, since 2016, been on a mission to militarily subdue groups of armed civilians seeking to (re-)establish the autonomy of British Southern Cameroons–that is, what has been the country's two minority English-speaking regions under different governance structures since 1961. Apart from the millions of human lives they claimed, such conflicts inflict a lot of damage on urban amenities, including green infrastructure. In Sierra Leone, for instance, urban water resources were destroyed while landmines rendered parks and other green areas in built space unusable for extended periods. In the English-speaking regions of Cameroon, hundreds of villages, including valuable pieces of green infrastructure have been completely burnt.

Socio-economic and Demographic Context A glance at global demographic statistics reveal that Africa has one of most youthful populations in the world. The average age of the continent's population is 19.5 with 60% younger than 35 years (World Facts 2019). The sub-Saharan region is largely sparsely populated especially in hinterland areas. The density of the region ranges from a low of 3 persons per square kilometer in Namibia to high of 624 persons per square kilometer for Mauritius. However, it must be noted that the density of Mauritius is an outlier as most densities throughout the region hover within the two-digit range. On the whole, the region is projected to experience significant growth within the foreseeable future. Typically comprising rural inhabitants, the region has witnessed, and will continue to witness, significant growth of its urban population. Slum settlements are expected to experience both physical and demographic growth. This expansion will invariably put pressure and strain on natural resources, including food, energy and water within urban areas and their peripheries. Rapid population growth and the commensurate expansion of built space is likely to cause many problems. These have implications for natural resource availability and use in built space in general and urban areas in particular. These trends have been known to strain sanitary facilities, compound waste management problems and lead to inappropriate sanitary behavior (Global trends 2019).

A number of economic trends with implications for natural resource availability and use in built space are also worth noting (Schneidman et al. 2019). For a long time, especially during the colonial era, the sub-Saharan region seldom felt the effects of positive trends in its economy. As colonies of economic exploitation as opposed to settler colonies, all economic gains from territories in the region were transmitted to the colonial master nations. Thus, at that time, the urban centers experienced physical and demographic expansion but very little economic development. This trend remained largely the same subsequent to the demise of colonialism, thanks to the fact that economies in the region continue to depend on those of their erstwhile colonial master nations. Despite this, the region's economies continue to experience steady growth; they are projected to grow by 3.8% making the region one of the fastest growing in the world in 2019 (Schneidman et al. 2019). The collective growth for the region's three largest economies, namely Angola, Nigeria, and South Africa is projected to be 2.5%. A growth rate of 5% over the next 5 years is projected for some countries in the region, including Burkina Faso, Tanzania, Uganda, Kenya, Senegal, Benin, Côte d'Ivoire, Ethiopia, Ghana, and Rwanda.

Ecological Context The sub-Saharan African region sits astride the Equator, hence, the appellation, Equatorial Africa in reference to the region's middle belt. It contains lush forests along the equatorial belt, and grass lands within the southern reaches of the Sahara Desert and Kalahari Desert area. Freshwater is plentiful within the equatorial zone; meanwhile, the Sahel region suffers from problems of water scarcity. Natural disasters, including droughts, floods and degradation of soil fertility are commonplace in the region. The Sahara Desert has been expanding southward over the years. This, coupled with pressures from human activities, grazing and logging are contributing to rapid deforestation and biodiversity loss in the region. By some estimates, the region is experiencing a deforestation rate twice as fast as that of the world at large. This situation has been aggravated by unregulated human activity such as uncontrolled burning, wood and charcoal cooking.

3.3 Culture and Environmental Stewardship in Sub-Saharan Africa

People's perception of, and orientation towards, the natural environment as well as the way they view themselves vis-à-vis nature in general is a function of their culture. A fundamental element of this culture is religion, which may be defined to include (Njoh 2006: 32):

beliefs, feelings, doctrines and practices linking a people (usually a community of believers) to a sacred and higher level being.

The indigenous religions of the sub-Saharan African region have been collectively referred to as African traditional religion. This should not be taken to mean that there is a common set of religious doctrines throughout the region. Rather, there are sets of beliefs shared by people of the same tribal origin or ancestry. For example, Yoruba religion is mainly for people who trace their ancestry to southwestern Nigeria, while adherents of Igbo religion comprise people of southeastern Nigerian extraction. Similarly, the Zulu religion is practiced by persons who are indigenous to southern Africa. However, while there may be no universal African religion, the named and other African religions share at least five features. The first, and most important, for the purpose of this book, is the fact that all of the religions see people as creatures on par with other creatures of the Almighty God. Thus, adherents of indigenous African religions do not, as Judeo-Christians and Muslims believe, that people have dominion over other creatures. Second, adherents to all indigenous African religions treat religion—that is, their beliefs and commensurate practices—as a way of life. In other words, religion cannot be separated from other aspects of livelihood. Thus, within the ethos of indigenous African religion, everything in the universe is inextricably interconnected. Third, all indigenous African religions harbor a holistic view of the universe in which the body and its social and natural environments are viewed as spiritually interlinked. Within this holistic framework, an ailment or sickness affecting any part of a system transmits reverberating signals throughout that system. Thus, illness within indigenous African religious ethos is interpreted as a function of an imbalance not only of the body but also the person's social life as well as natural environment. Such an illness may be attributed to the patient's failure to treat the natural environment with the care it deserves. Fourth, all African indigenous religions treat land with reverence, and consider it one of the Almighty God's most precious creations. Accordingly, adherents to these religions believe that no one owns or can own land; therefore, it cannot be sold (Njoh 2006). Rather, people have only use—and not ownership—rights over land, which is bequeathed to those alive by the dead to be carefully used and placed at the disposal of the unborn. Therefore, those alive are nothing more than custodians of land; in this capacity, they have the duty of maintaining and preserving land. It is essentially for this reason that traditional Africans are wont to frown at efforts by Western change agents to commodify land. Fifth, the religions are unified in treating bodies of water including springs, rivers, streams, the sea and oceans with reverence. In fact, mythologies with water at their core are commonplace throughout Africa and parts of the African diaspora. The indigenous religions of West, Central, East, Southern Africa and some parts of the African diaspora possess and venerate a water deity commonly known in pidgin English, the lingua franca of Anglophone West Africa, as Mami Water. Reverence for Mami Water meant that traditional Africans could not indulge in any activity deemed injurious to any body of water. This is because doing so would provoke the wrath of Mami Water.

Indigenous African religion is also well-known for promoting the creation and preservation of green infrastructure in built space. An example of such infrastructure is the sacred groves that constitute a conspicuous feature of built space among the Meta of Cameroon (Njoh 2006) (also, see Box 3.1 below). To be sure, sacred groves as a feature of the built and natural environments are neither new nor unique to the Meta's. This is despite the misleading accounts of Western scholarship that are wont to treat the phenomenon as cultural relics from Africa's 'static precolonial past' (Sheridan 2009). This tendency, some have suggested, is a function of the fact that the religious purpose and function of these groves are often overemphasized (e.g., Sheridan 2009). By so doing, their social and ecological values are overlooked.

3.4 Land in Built Space

Land is arguably the most important factor in the discourse on natural resources in built space. It is both a natural resource in its own right and the most significant determinant of the quantity and quality of other natural resources available in built space. Thus, how land is controlled, and who controls it—in other words, land tenure—determines how much of it can be set aside for the purpose of promoting nature in built space. This accentuates the importance of land tenure systems. In the case of Sub-Saharan African countries, these systems are firmly grounded in Africa's triple heritage actualized in terms of indigenous African values, and the exogenous influences of Islam and European colonialism/Judeo-Christianity.

Elsewhere (Njoh 1998), I have retraced the evolution of land tenure policy with emphasis on the impact of exogenous and endogenous forces in Africa. Indigenous or pre-colonial sub-Saharan Africa had no written legislation governing the relation between humans and land. This should not be misconstrued to suggest that there were no principles governing land use and control in the region prior to the European conquest. Indeed, these principles existed and had at their core, the belief that land was a gift from God that could neither be owned nor sold. Thus, the relationship between humans and land as enshrined in indigenous African tradition is simply one of trustee with usufructuary rights. However, people had the privilege to use land on a 'right-of-first-occupancy' basis. Implicit in the notion of occupancy of any parcel of land is the effective use of that parcel. This invokes the ability to clear the land or put it to some effective use such as hunting (Selase et al. 2015). Thus, the first family to occupy a piece of land had the privilege to control and use it. The units that were endowed with the legitimate right to occupy, control and use land are: a family, comprising a man, his wife or wives and children; an extended family, that is, two or more families that have come together to constitute a hamlet; and a village, which is a human settlement of two or more hamlets. All parcels of land not under effective occupation or use was considered communal lands under the control of the village chief. Also known as stool lands, they are available for use to benefit the community at large. To the extent that public parks, and green infrastructure in general, benefit the community at large, they exemplify such use. It is therefore safe to conclude that African traditional land tenure systems contain provisions that can facilitate efforts to promote nature in built space.

Islamic Influence The onset of this influence dates back to the thirteenth century when Arabs began settling in Africa in significant numbers. A prominent element in the cultural baggage that Arabs brought to the continent is Islam. Here I hasten to note that Islam predated Christianity in most parts of the continent by about a millennium. The arrival of Arabs in Africa was accompanied by the introduction of new social and economic relationships between societies and their natural resources. To be sure, some features of Islamic and Indigenous African beliefs regarding humans and their relationship to land coincided. For example, the concept of *hema* (plural *ahmia*), a system for conserving and managing communal land in Islamic traditions bears a striking resemblance to the sacred groves in indigenous African culture. However, once in Africa, the Islamic leadership did nothing to identify the common threads that run through Islam and African traditional beliefs. Instead, they strived fervently to supplant these beliefs with Islamic doctrine. Thus, the earliest efforts at codifying the principles of land rights which occurred under Ottoman Rule in the Middle East in 1858 were contemporaneously transferred to Sub-Saharan Africa. These efforts divided land into two main categories, namely *miri* or *amiri*, that is, property of the Amir; and private land or *Mulk* (Rae, n.d.). Private land was further divided into two categories, *raqabah* (concerned with resource ownership), and *tassaruf* (focusing on usufructuary rights. Land in the first of these categories falls under the orbit of the state, while land in the second comes under the sphere of the land user(s). This marked the earliest formal introduction of the tripartite notion of land as property of the state, land as private property, and land as a commodity in many parts of Africa. It also marked the point at which the burden of conserving land in built space first became the responsibility of the state throughout the continent. This is in contrast to the indigenous African tradition which placed this responsibility in the hands of individual families or their local communities at large.

Land commodification is the feature of Eurocentric land reform initiatives that has proved most consequential for nature in built space in Africa. The feature bodes particularly ill for urban minorities and the poor who must often take up residence in slums. These are, by definition, overcrowded, shoddily constructed structures perched on the most precarious patches of land in the least desirable parts of periurban districts. A conspicuous feature of slums throughout Africa, and elsewhere for that matter, is the complete lack of green infrastructure. The case of Kibera, Nairobi, dubbed Africa's largest and most densely populated slum is illustrative (see Fig. 3.2).

As can be gleaned from the photograph of Kibera shown here as Fig. 3.2, there is no land for farming and recreation. In addition, there is hardly any tree or any vegetation. The lack of agriculture, green and similar pieces of infrastructure means that slum dwellers cannot take advantage of the economic opportunities offered by nature in built space. It is well-established that agricultural activities and green infrastructure in general can create employment opportunities, generate revenue, enhance aesthetic appeal, prevent soil erosion and control climate. However, none

Fig. 3.2 Kibera Slum, Nairobi, Kenya. (Source: commons.wikimedia.org. https://commons.wikimedia.org/w/index.php?search=Kibera+slum&title=Special%3ASearch&go=Go&ns0=1&ns6=1&ns12=1&ns14=1&ns100=1&ns106=1#/media/File:Kibera_Slum_-_panoramio.jpg)

of these advantages is likely to accrue to slums or their residents. Thus, it is safe to conclude that, due to their location, residents of slums such as Kibera are victims not only of social but also spatial injustice.

The genre of injustice in question can be best understood in the context of the ongoing discourse on the right to the city (Njoh 2015; Soja 2010; Mitchell 2003; Harvey 2008). Drawing inspiration from the earlier works of Henri Lefebvre on issues such as the production of space *(la production de l'espace)* (Lefebvre 1974), and the right to the city *(le droit à la ville)* (Lefebvre 1968), works on this theme incriminate capitalist and peri-capitalist societies as sites of expropriation by a dominant class.

3.5 Water Resources and Management

Water occupied a crucial place in the lives of pre-colonial sub-Saharan Africans. However, water was not evenly distributed across the region; it was/is plentiful around the Equator but scarce in areas around the desert—the Sahara to north and the Kalahari to the south. In general, indigenous Africans, regardless of their location, had developed water principles, philosophies, management strategies, and technologies that would be considered sophisticated even by contemporary standards. For example, recognition of the indispensability of water in human life, and the fact that unlike air, it is location-specific, led to the development of complex rules governing its access and use. In this connection, water, more than other resources, was considered a communal resource that had to be managed by a community leader for the general good of the community as a whole. There is also evidence attesting to the fact that ancient Africans were fully aware of the inextricable link between water and sanitation. The work of Mbatha et al. (n.d.) has uncovered some of this evidence in the case of the Zulu's in present-day South Africa. The work highlights philosophies and doctrines on human-water relations that are sophisticated even by contemporary standards. For instance, rivers were good for washing and bathing but not good enough for drinking. From this, it is reasonable to conclude that recognition of the importance of water to attain hygiene and sanitation goals date back to antiquity in Africa. In fact, as Mbatha et al. (n.d.) observed, ancient Africans are known to have enacted and implemented laws that mandated the appropriate disposal of solid wastes. The earliest method for human waste disposal was the cat method. This consists of digging a shallow hole in the ground, excreting and burying the excreta in the hole. Excreting in any body of water, whether still or flowing, was considered a taboo. There were no restrictions on the disposal of human liquid waste. Frequent hand-washing, but especially after the use of toilets, was a requirement. To facilitate this, a clay basin was placed by the door of people's huts. Ancient Africans were also well-known for using water for curative purposes. For instance, they used water for hydrotherapy, which involved the use of cold, hot, compressed and steamed water vapor for curative purposes (Emeagweli and Dei 2014; Njoh 2006). In addition, they demonstrated acute knowledge of the value of clean drinking water for human health; they also instituted basic rules that ensured the protection and maintenance of drinking water sources, including springs and rivers (Njoh 2006).

Ancient Africans also demonstrated a deep understanding of water sources, including rivers, lakes, springs, aquifers, and shallow and deep wells. In addition, they also possessed knowledge of the annual hydrological cycle, the characteristics of water systems such as the natural flow, topography, water sources including water bearing plants and ideal locations for shallow wells. With the exception of rivers and lakes, all sources of fresh were considered good for drinking. There is also evidence that ancient Africans had developed functional technologies for the conveyance and transportation of water. Prominent among these technologies were ostrich eggshells, animal bladders and gourds. These were used to fetch, transport and store water. In other parts of the sub-Saharan region, calabashes, clay pots, and animal horns served the same purpose.

Three factors nullified the need to develop more sophisticated water distribution and management technologies (Mbatha et al. n.d). The first was low population density in built space. The second was the lack of conflict, and the third was the wide range of location choices available to ancient Africans. In what is present-day South Africa, the Zulus are known to have preferred living in areas with an abundance of water resources. With the arrival of Europeans in the region, and especially with the onset of the colonial era, colonial authorities found these water-resource-rich areas fertile ground for commercial agriculture. Consequently, they proceeded to forcibly remove members of the indigenous population from these areas. In their stead, they located European farmers in these water-rich areas. Members of the indigenous population were decanted into less desirable areas characterized by water scarcity problems. This essentially marked the beginning of the pattern of racial spatial segregation that was the hallmark of South Africa's apartheid system. This spatial configuration pattern was later adopted by European colonial authorities; it constituted the basis for water service provisioning throughout the sub-Saharan African (SSA) region.

The European colonial era formally began in SSA in 1884 although it must be noted that European presence, complete with exclusive European enclaves, in the region predated this time. Hardly any issue was as dominant as water in the discourse on natural resource management in built space in the region then. Although colonial authorities ensured the location of European settlements in areas richly endowed with water resources, they faced other seemingly intractable problems. One of these was making water available where and when it was most needed. The earliest and best-known efforts in this connection involved a heavy dose of forced manual labor input from Africans. For instance, colonial authorities in Entebbe, Uganda, operated water schemes that required Africans to fill water tanks in exclusively European and Asian districts with water from the rain and spring sources (Nilsson 2017). Similarly, as I noted elsewhere, Africans were required to fill water tanks perched atop roofs in the European district in Kumbo, colonial Cameroon (Njoh 2006).

Thus, the earliest European efforts to manage water in built space in sub-Saharan Africa were quite rudimentary and mirrored the indigenous ones they worked hard to supplant. In this regard, Europeans, like Africans depended on rain, springs, rivers, lakes, and shallow wells to address water needs in built space. However, with rising population densities, shallow wells soon proved unsafe. For instance, all of the public wells developed in Lagos in 1882 under the auspices of the governor of Lagos colony, Sir John Glover (1829–1885), were deemed unsafe by the Colonial Surgeon General. A laboratory test of water samples from wells in the town had revealed the presence of contaminants from sewage and cesspools. To prevent disease and deaths resulting from this and other contamination, the surgeon prescribed a 50-foot distance between a water well and the closest residential facility in colonial Lagos. Additional safety measures that were ratified include the following requirements: the clay- or cement-padding of wells, the use of well pumps instead of bucket to extract water from wells, and the covering of wells when not in use. In 1910, almost three decades after Governor Glover launched the project to develop public wells in Lagos, the city witnessed the completion of its first modern water supply system. The project, which was one of the first in colonial sub-Saharan Africa, was commissioned on July 1st 1915 by Nigeria's colonial Governor General of the time, Sir Frederick Lugard.

A notable feature of the Lagos water supply project, like others elsewhere in the region during the colonial era, is the variable coverage; this ostensibly favored exclusive European enclaves over African areas (Hungerford and Smiley 2016; Njoh and Akiwumi 2016). This racialized infrastructure and natural resource distribution system ensured that European enclaves were furnished with water piped to individual households. At best, sparsely distributed stand pipes served African neighborhoods, and in most cases, these neighborhoods were left to their own devices with respect to meeting their water needs. This was the case throughout the region irrespective of the colonizer's nationality. The recent work by Hungerford and Smiley (2016) comparing the water infrastructure development and distribution policies of British and French colonial authorities lends credence to this assertion. Their analysis of Dar-es-Salaam's colonial water network is particularly revelatory as it provides further evidence of the universality of racial segregation as a crucial element of European colonialism. The Germans ceded control over Tanganyika of which Dar-es-Salaam was the colonial capital, to the British as part of the outcome of the conclusion of World War I. The British inherited a segregated urban structure that divided urban residents into three distinct racial spaces for Europeans, Indians and Africans. The exceedingly limited water network served exclusively the European/colonial administrative districts while small portions of the rest of the town were served by shared water kiosks. This situation, characterized by a variable distribution of water services, prevailed in Dar-es-Salaam throughout the British colonial era in Tanganyika (present-day, Tanzania). In Kenya, another British East African colony of that time, colonial authorities operated similar racialized water distribution systems. Nairobi, which rose to prominence in the colonial theatre as the first campsite for workers and headquarters of the Uganda Railway Corporation soon became the seat of the colonial government. The town faced severe water problems that became increasingly intense over the years. The initial piped water network for the town was developed by the Railway Corporation and designed to serve the railway station's European districts. In the 1920s, British colonial authorities took over control of Nairobi and maintained water schemes that ostensibly favored European enclaves. By 1948, and as revealed in the city's Master Plan, Nairobi's water supply network, which tapped water from River Nairobi, had become woefully inadequate that it could meet no more than one-sixth of the city's water needs (Hungerford and Smiley 2016). Yet, because of the racially segregated structure of the city, and the bias in favor European districts, only the non-European areas of the city felt the acrid pinch of the water insufficiency problem. To shed more light on the biased nature of colonial Nairobi's racialized water policy, Hungerford and Smiley (2016: 79) made the following observation. At some point during the colonial era, the non-European areas received only 45 liters of water per day compared to the daily supply of 220 liters that went to the European enclaves.

French colonial authorities are also on record for racializing the provisioning of natural resources such as water in built space. Thus, the European enclaves in French colonial towns enjoyed qualitatively and quantitatively higher levels of water access than non-European districts. For instance, the piped water system that French colonial authorities installed in St. Louis, Senegal in 1886 served exclusively the European districts (Njoh 2007). A similar situation prevailed in French colonial towns such as Niamey, Niger; Dakar, Senegal; Brazzaville, Congo (PR), and Conakry, Guinea. Each of these towns had an exclusive European enclave, which the French called *'le Plateau.'* The selection of sites for colonial towns took advantage of yet another natural feature in built space, namely elevation. Elevated sites were deemed important because they were invariably more airy, ventilated and elevated than surrounding areas. The elevation was supposed to provide the colonial authorities with an uninterrupted view of the indigenous districts in the low-lying areas. When it came to the provisioning of water services, these exclusive European enclaves were prioritized over the indigenous districts.

Elsewhere (Njoh 2011: 3), I noted the consequential fate of the European enclaves subsequent to the demise of colonialism and the commensurate exit of Europeans.

> [Once the Europeans left] the indigenous leadership assumed control of the colonial government offices, and re-assigned the erstwhile Eupean quarters to members of the upper echelons of what at the time was an emerging independent government bureaucracy. At the same time, they designated the erstwhile native areas as the residential district of low-income members of the society. While the upper-income areas continued to benefit from some degree of upkeep and maintenance, particularly in terms of utility provisioning and planning code enforcement, the low-income areas were virtually neglected. With the passage of time, the low-income areas have grown to unprecedented levels as deteriorating conditions in rural areas increasingly result in mass rural-to-urban migration.

A notable upshot of the afore-characterized development is the effective transformation of racially segregated, into socio-economically compartmentalized space. Thus, the variable water provisioning schemes that favored exclusive European enclaves during the colonial era was re-fashioned to favor bureaucratic and business elites following the demise of colonialism. However, it is necessary to note that very little has been added to the stock of public infrastructure, including water facilities inherited from the colonial era in sub-Saharan countries (Hungerford and Smiley 2016; Njoh and Akiwumi 2011). This means, among other things that, the heightened rates of urbanization that have occurred since the countries gained political independence have not been matched by an increase in the inventory of water infrastructure. This explains the gross deficiency in access to water that prevails in the sub-Saharan African region. However, as Njoh and Akiwumi (2011: 155) have observed, the region manifests extremities in this regard. While some parts of the region—such as those around the Congo, Nile and Zambezi rivers and Lake Victoria-Nyanza—are blessed with an abundance of water, others—such as those in the Sahelian belt—are not. In any case, the supply of water to built areas in sub-Saharan Africa has always been problematic with the most nagging problem being financial resource scarcity. Accordingly, the most successful water projects in the region are those that have relied on a time-tested strategy, namely community participation (CP). Considered a bonafide indigenous African project development and management strategy, CP maximizes citizen in-kind input while minimizes dependence on financial resources. As I noted in a previous study, the residents of Kumbo in the N.W. Region, Cameroon employed CP during the construction and post-construction phases of their community water project (see Njoh 2006). In the construction phase, CP entailed citizens contributing in the form of labor and cash based on their sex and income levels. In the management phase, CP has entailed groups of citizens taking turns cleaning and maintaining communal water fountains and their immediate surroundings.

3.6 Afforestation and Greenery in Built Space

Natural Resources in Built Space in Traditional Africa Indigenous African polities operated three main types of land regimes based on societal units in built space. Societal units include: the immediate family, the extended family and the village. Thus, there are immediate family-controlled, extended family-controlled and village or regional communally-controlled lands. Each of these units is required to set aside some parcels of the land under its controls for conservation purposes or as green space for recreational, ceremonial and other uses. Box 3.1 describes three such parcels and their corresponding green uses in Meta, Cameroon. It must be noted that, over the years, Metaland has grown demographically resulting in a significant increase in population density. Consequently, and as elsewhere throughout Africa, there has been a significant decline in the number and overall area of green spaces in indigenous built areas.

Box 3.1 Green Areas in Built Spaces of the Meta, Cameroon

The need to preserve nature in built space is enshrined in indigenous African religious doctrine. Three types of such spaces that have been preserved by the Meta, an acephalous ethnic group on the Bamenda Plateau in Cameroon's North-West Region are described here for illustrative purposes. They include, the family compound farm, the extended family ceremonial field *(sami),* and the sacred grove. The institutional structure of the Meta contains three main levels, including the immediate consanguine family headed by the family head, the extended family headed by a quarter head, and the village headed by the village chief.

The immediate family and family compound farm. The immediate family includes a man, his wife or wives and children within a family compound. The family compound typically measures about two-tenths to six-tenths of a hectare (Njoh 1999; Dillon 1990). It always contains a main house for receiving guests, the house of the family head and one house for each wife and her children; and more importantly, a family farm—the family compound farm. This typically includes fenced areas containing livestock such as pigs, goats, sheep and chickens. Meta people believe that the Almighty God occasionally pays unannounced visits to his creations, particularly human beings. God, according to Meta religious thought, may assume the form and identity of a regular human being and appear at any time of the night or day. The family compound farm is thus designed to serve as a source of food for the unannounced visitor who may be an incarnation of God.

The extended family and hamlet ceremonial field. The Meta extended family typically lives in small hamlets. Each of these is a congregation of family compounds, which share a common ceremonial field known in the Meta lan-

(continued)

Box 3.1 (continued)

guage as *sami*. This is used for extended family ceremonies such as the celebration of births, marriages and funerals. Similar occasions are conducted in the Village *Sami* when the parties involved are of a higher social standing.

The village and village ceremonial field. The Meta polity is organized according to villages of which there are 31. Each of these is headed by a chief who lives in a palace of several houses congregated around a large courtyard to form a rectangle or a sphere. Adjacent to the palace is a field which may be as large as a soccer pitch. The field is typically used as the venue for celebrating occasions of significance to the village as a whole, including but not limited to the beginning of the harvest or planting season, the death of a village luminary, and the sharing of a large animal such as an elephant killed for food.

The sacred grove. The third notable natural feature in built space as stipulated in Meta traditional religious doctrine is the sacred grove. Sacred groves, the Meta believe, provide a quiet, serene and unadulterated habitat for ancestral spirits. Meta people often invoke the spirits of their ancestors to help resolve obstinate and intractable problems.

Source: Author's first-hand research/experience.

One piece of green space that has withstood the pressures of urbanization in Africa—perhaps because of its ritualistic value—is the sacred forest grove. Thus, despite the increasingly degrading landscape of built space in Africa, sacred groves continue to enrich the continent's biodiversity. In Ghana, for instance, the government has crafted innovative programs founded on the reverence for sacred groves to promote their preservation (Amoako-Atta 1998).

Influence of Colonialism/Christianity on Natural Resources in Built Space European colonialism stamped some of its most indelible marks on built space in Africa. Here, we hasten to note that the European colonial era on the continent coincided with the emergence of modernist urban planning as a profession in Western Europe. However, Western Europe lacked the genre of physical space that was necessary to test the lofty plans that emanated from the drawing boards of town planners of the time. Consequently, the vast expanse of Africa provided just the type of physical space they needed to test the workability of human settlement plans especially those concerned with spatial organization. The most notable features of these plans for the purpose of this chapter are: comprehensiveness, functional segregation and racial spatial segregation. Comprehensiveness is epitomized by Ebenezer Howard (1850–1928) in his Garden City plan. The plan as enunciated in Howard's classic, *To-Morrow: A Peaceful Path to Real Reform* (1898), which was revised in 1902 under the title, *Garden Cities of To-morrow*, sketched the blueprint for a city harmoniously co-habited by humans and nature. Specific features of the plan include the following (Njoh 1999): a 5000–6000-acre tract of land of which only a small portion is occupied by buildings and ancillary infrastructure built around a garden, and contain at most 30,000 to 32,000 people. The rest of the land would be set aside for agricultural and recreational use. At about the same time that Western European countries were embracing the garden city model, colonial powers were adopting it as a template for their capital cities in Africa. For example, S.D. Adshead's 1931 plan for Lusaka, the capital of British colonial Northern Rhodesia (present-day, Zambia) employed the garden city template. On their part, the French, adopted the model or what they call, *cité-jardin*, as the design of choice for cities in their dependencies such as Morocco, Tunisia and Algeria.

Modernist planning, of which the Garden City model is a part, emerged in Western Europe as a formal response to the negative externalities of the industrial revolution of the eighteenth Century. Once transplanted to Africa, colonial authorities meticulously employed it to actualize their racial residential segregation desires. Racial residential segregation policies resulted in two main types of residential areas—one with a high density and almost no green infrastructure, and the other with well-manicured lawns and rich natural vegetation. The latter, the exclusive European enclave, was in some cases such as Freetown, Sierra Leone, often located atop a hill overlooking the residential areas of the 'natives' at the foot of the hill. In all cases, the European enclave was separated from the areas of the 'natives' by a greenbelt. Thus, while the residential areas of Africans contained hardly any formally developed green infrastructure, the European enclaves boasted well-manicured lawns. Colonial authorities also introduced many social activities, particularly Eurocentric outdoor sports such as soccer, golf, tennis, netball, and volleyball that required purposefully designed green infrastructure. To be sure, this infrastructure had different objectives, such as assimilation and acculturation, in the context of the colonial project. However, there is no question that it went a good way in augmenting the quantity of green space in built areas throughout the Sub-Saharan African region.

Other colonial era facilities that resulted in augmenting the stock of green infrastructure in built space in sub-Saharan Africa (SSA) are botanical and zoological gardens. These were initially designed for the purpose of promoting colonial scientific research objectives, especially with respect to knowledge of tropical plants and animals. Currently, these facilities play an important role in the preservation of endangered plant species and wildlife. Additionally, they serve as a locale for stress relief, and provide a sense of place to the indigenous populations of their respective locales. Their role in biodiversity preservation is illustrated by the Botanical Gardens in Limbe, Cameroon (Fig. 3.2). This facility was established by German colonial authorities in 1892 in what was at the time, the small town of Victoria. Today, the Gardens literally constitute a lush forest situated in the increasingly urbanizing town, which is now a fully blossomed city and had since been renamed, Limbe. The last colonial era piece of green infrastructure in built space that was inherited by indigenous authorities and remains a conspicuous part of the urban landscape in SSA are street trees and vegetation. Streets and roads flanked by these are becoming ubiquitous features of cities undergoing a face-lift in many parts of the region. This is true even in a country such as Ethiopia, which never experienced any sustained form of European colonialism. For instance, as shown in Fig. 3.3, tree/vegetation-lined streets are a common sight in the country's second largest city, Mekelle.

3.7 Agriculture and Food Security in Built Space

Agriculture in built space in sub-Saharan Africa can be better understood within the broader context of the region's agricultural experience as a whole. Paradoxically, while the region constitutes the cradle of homo sapiens, it is not the birth place of agriculture. This credit belongs to the Fertile Crescent of the Mesopotamia and Tigris rivers, where agriculture originated in 9000 BCE. Some two millennia later in 7000 BCE, people in China and New Guinea also took up the practice. Due to their proximity to, and interaction with Southwest Asia, Egyptians also followed in the footsteps of people in West Asia. They began practicing agriculture a considerable while before sub-Saharan Africans. It was not until 3000 BCE that the first agricultural activity was recorded in sub-Saharan Africa, precisely within the Savannah region, between present-day Cameroon and Nigeria. This development predated organized community living, hence, the built environment writ large. However, once Africans began living in communities, they embarked on crafting strategies for meeting their food need at minimal cost in terms of the time and distance required to reach food sources. Some of these strategies have been resilient enough to survive centuries of the assault of many exogenous forces. Box 3.2 describes a strategy that has been used by the Shona's and Ndebele's in Zimbabwe, to ensure availability of food in times of poor harvest or natural disasters.

Box 3.2 The Practice of Zunde Ramambo (Among the Shona) or Isiphala Senkosi (Among the Ndebele) in Zimbabwe

Zunde ramambo and Isiphala senkosi mean "the chief's grain reserves." this time-proven indigenous system to guarantee food security, entailed designating a piece of communal land as the community's farm. This is then cultivated jointly by, and for, members of the community at large. Food from this farm is stored in silos at the chief's palace. From there, the chief would dole out food to households in need. The typical beneficiaries of food from these silos are widows, orphans, and stranded strangers. During times of poor harvest or famine, all members of the community would turn to the Chief's reserves for food. The decisions regarding the quantity of food required to be saved or distributed involved not only the chief but his advisers and other community notables.

Source: Swikepi, C. (2011). Community Participation and Food Security in Zimbabwe: The Case of Marange Area in Mutare District. Ph.D. Dissertation, University of Fort Hare.

Kuo, L. (2016). A 700-Year Old West African Farming Practice Could be an Answer to Climate Change. https://qz.com/africa/713512/a-700-year-old-west-african-farming-practice-could-be-an-answer-to-climate-change/

Apart from developing strategies to reduce the time and effort to reach food sources, Africans also crafted techniques to enhance agricultural productivity. One of these, depended directly on kitchen waste, particularly wood ash and charcoal. This waste is typically used as manure to enhance soil fertility. Soil treated with this waste is what has been called 'African dark earths' (Kuo 2016). Sub-Saharan Africans have employed 'African dark earths' in their practice of agriculture in built space for over seven centuries. This is because they recognize that soil in built space lack the quantity and quality of vegetation available in the natural environment.

The colonial era did very little to promote urban agriculture in Africa. If anything, colonialism, because of its emphasis on economic productivity, fiercely promoted cash crop/plantation farming at the expense of food crop/subsistence farming. In addition, colonial authorities either discouraged urban living for Africans or heaped them in densely populated precarious locales that had no space for farming. In contrast, the exclusive European enclaves were located in sparsely populated hilltops or plateaus. The luxurious residential facilities in such locales boasted large lots, low floor-area-ratio (FAR) and ample room for large well manicured lawns, gardens and orchards. Such facilities belonged in most cases to the colonial government; in a few cases, they were properties of European business enterprises such as mining or plantation agriculture corporations. The care and maintenance of the facilities, including the residential units, lawns, orchards and gardens was not the responsibility of the tenants. Rather, colonial governments and private corporations engaged personnel who attended to their respective facilities on a full-time basis. Thus, the only planned and formally recognized urban gardens that existed during the colonial era in sub-Saharan Africa were in the exclusive European enclaves.

The 1960s witnessed widespread decolonization in the sub-Saharan African region. As mentioned earlier, one of the outcomes of the decolonization process is the fact that senior members of the independent government and business elites took over the erstwhile European enclaves. As for the rest of the urban areas, they became the destination of choice for rural residents in search of the illusive dream of a better life. This led to alarming rates of population growth, increased densities and overcrowding in urban areas throughout these countries. Most of these demographic changes occurred in the low-income areas. Another outcome of the decolonization process was a tendency on the part of the indigenous leadership to enact spatial regulation policies that were more stringent than those they inherited from their colonial predecessors. For instance, authorities in Dar-es-Salaam forbid the growing of crops within 14 meters of roads or within 15 meters of river banks (Beach 2013). Apart from institutional constraints such as this, the urban poor in sub-Saharan Africa face a number of other barriers in their bid to practice urban agriculture. Prominent among these are: lack of land due to the overcrowded nature of low-income areas; water scarcity, which is a ubiquitous problem in low-income areas; and insecurity, or lack, of tenure over land, which often leads to the frequent eviction of the poor from land they occupy.

Despite these barriers, urban agriculture remains an important pastime and/or vocation in sub-Saharan African countries. The observation of Beach (2013) is very informative in this regard. In the Zambian capital, Lusaka, more than half of the households practice one or another form of urban agriculture. In cities such as Kampala, Uganda, and Douala and Yaounde, Cameroon, many households raise livestock including poultry, dairy, and pigs. Urban farms in these countries assume many forms, including vegetable and small animal farming in backyards or undeveloped plots of land in towns and cities. In hinterland or non-coastal areas, urban agriculture often includes one more activity, namely, fish ponds. Farms in built space, particularly in sub-Saharan African countries are typically located along roadsides, rivers, under power lines, at the urban periphery, on institutional land (e.g., land belonging to educational and administrative institutions).

I would be remiss if I ended this discussion of urban agriculture in sub-Saharan Africa without acknowledging its economic capabilities. It is a dependable source of income for many. For example, Beach (2013) observed that in Bamako, Mali and Dar-es-Salaam, Tanzania for instance, urban farmers make as much money from selling surplus food from their farms as unskilled construction workers earn each month. She further observed that people with farms in Yaounde, Cameroon's capital city, earn 50% above the local minimum wage from sales of surpluses from these farms. Further evidence attesting to the economic importance of urban agriculture has been uncovered by other researchers, including Mireri (2002) with respect to animal husbandry, particularly pig rearing, in Nairobi, Kenya; Gockowski et al. (2003) and Danso et al. (2002) in the case of vegetable farming in Yaounde, Cameroon, and Kumasi, Ghana respectively. Further evidence of the enormous contribution of urban farming comes from Addis Ababa, the Ethiopian capital, where cattle reared within the city limit—or so-called, backyard farms—produced as many as 20 million liters of non-pasteurized milk per day in 2000 (Tegegne et al. 2000).

3.8 Energy in Built Space

Energy is a bona fide subject of interest in the discourse on built space; if for no other reason, because its end users are almost always located in built space. Like many other facets of its profile, the history of energy provisioning in sub-Saharan Africa has not been adequately researched. Therefore, any meaningful effort to analyze the region's energy generation initiatives and their implications must perforce examine this history. The history is intimately tied to the region's colonial past. Yet, it would be erroneous to state that energy generation and consumption were unknown in the region prior to the European conquest. A few recent studies have documented energy generation initiatives in the region predating the colonial era (see e.g., Mavhunga and Trischier 2014; Showers 2011; Kanduza 2009; Phillip 2006). Far fewer studies have attempted to address questions such as 'why' and 'how' these initiatives emerged. How, for instance did pre-colonial Africans generate energy? Why did they do it? Our aim in this section is to tackle this perennial question and show how European colonization established the foundation for modern electricity generation and consumption in the built environment in sub-Saharan Africa.

The literature on the history of energy generation and consumption in Africa is replete with gaps and erroneous accounts. Most of the accounts err by tying this history to colonialism (e.g., Showers 2011). In this case, they speciously claim that Africans knew nothing about energy generation and consumption prior to the arrival of Europeans on the continent. Such accounts ignore the significant achievements of Africans in the development of built space and commensurate infrastructure. Yet, history contains copious accounts of these achievements not only in Africa but also beyond. Consider the case of the Moors, the Black African ethnic group that conquered Spain in 711 AD (Black History Studies Online). During the heydays of their occupation of Spain, the Moors transformed Cordova, the seat of their government, into the most modern city in Europe. The city boasted paved streets and raised sidewalks for pedestrians. More importantly for the purpose of the discussion in this section, the streets were well-lighted at night. This was long before streets in London and Paris were furnished with lights. Evidence of energy generation and consumption predating the European conquest of sub-Saharan Africa in particular also abounds. Some of this evidence can be gleaned from the fact that long before this conquest, sub-Saharan Africans were already actively involved in energy-dependent industrial activities. For instance, as Mavhunga and Trischier (2014) have noted, sub-Saharan Africans in the south-central region of Africa were mining and processing gold, iron and copper long before the arrival of Europeans. Similarly, Africans in the Ashanti area—roughly, the area occupied by present-day Ghana—operated highly functioning gold smelting plants that required refined energy to function. Additionally, archaeologists have uncovered remnants of mines as well as copper and gold smelting sites in the region, particularly in Karanda, Copper Queen, and Mupfure near Chegutu (Mavhunga 2014). However, it is fair to concede that the foundation for modern energy throughout Africa was laid by European colonial authorities. Energy constituted a critical element in the colonial project; it was necessary to realize the economic and socio-cultural objectives of this project. Accordingly, the earliest colonial electrical systems were designed to facilitate mining and agro-plantation activities in Africa in the first instance. In the second instance, electrical energy was necessary in exclusive European residential areas as part of efforts to replicate living conditions in vogue in Europe at the time. This was required to ensure the best living conditions possible for Europeans in Africa. The need to attain economic and social objectives of the colonial projects explains the rush to provide electricity to mines in Southern Africa as well as exclusive European enclaves in Kimberly and Cape Town in the late-1800s (Showers 2011). This was about the same time that George Denton, the Governor General of colonial Nigeria was seeking ways to address the need for street lights in the European enclave of Lagos (Olukoju 2012). Initially, colonial authorities employed gas lantern lights; these proved difficult to manage and were on from dusk until 11:00 pm (Olukoju 2012). The lighting question in Lagos was complicated by the fact that the city contained a sizable civil society comprising Western-educated Africans. These latter were intolerant of any urban development policies that were racially discriminatory by favoring exclusively European enclaves. Yet, when electric street lights were introduced in 1898, they, as their predecessors, the gas lantern lights that were introduced in 1891, were solely in the European District. This and other earlier efforts at electrification in sub-Saharan Africa employed mainly oil or gasoline-powered generators. With the passage of time, other sources of electrical energy, including hydro and thermal options, were introduced. An example of the latter is the Ijora Thermal Station, which colonial authorities developed to supply electricity to Lagos and proximate areas in 1923.

With the demise of colonialism throughout most of Africa in the 1960s, there was widespread adoption of ambitious electrification programs throughout the continent. In the sub-Saharan African region, many countries proceeded to procure loans from multi-national funding bodies such as the World Bank and the International Development Association to finance large-scale electrification projects (Marwah 2017). Most of the loans were intended to fund hydroelectric projects. The bias in favor of hydro sources was not accidental. Rather, it is a function of the abundance of water resources, particularly rivers in the region. However, the rivers are becoming increasingly less dependable as a source of electrical energy because of the rapidly dwindling volume of water they contain. That the rivers have been rapidly drying up in the recent past accentuates the need to judiciously manage the link between nature and built space. A leading reason for declining river volumes is deforestation and the loss of vegetation in general. The narrative in Box 3.3 articulates the link between forested land and water supply in general and in Africa in particular.

Despite the intense national and international attention directed at electrification projects subsequent to the demise of colonialism in sub-Saharan Africa (SSA), electricity remains a scarce commodity throughout the region. Africa as a whole accounts for more than a sixth of the world's population but generates only 4% of global electricity (Economist 2007, para., 1). As Table 3.1 shows, the problem is most intense in the SSA region. Here, access to electricity remains rare especially in the rural areas. Only 32% of urban residents and a minuscule 3% of rural residents in East Africa are connected to the national electricity grid (UN-Energy/Africa 2015: 68). The situation is worse in West Africa. Biomass in this sub-region represents as much as 90% of total energy consumption, and only 12% of the population

Box 3.3 The Link Between Forested Land and Water Supply

Two Russian scientists, Anastassia Makarieva and Victor Gorshkov (e.g., Makarieva and Gorshkov 2007, 2010) are at the forefront of efforts empirically linking forested land to freshwater supply. They have demonstrated that terrestrial vegetation cover helps to deliver moisture to land. They contend that because of their high leaf area index, natural forests are able to maintain evaporation fluxes involving transpiration and interception over the oceanic surface. Evaporation and transpiration on their part, jointly create an elevated

(continued)

Box 3.3 (continued)

moisture content in the vertical column of air over the forested land; this results in elevated condensation above the forest canopy due to the adiabatic ascent of moist air (Makarieva and Gorshkov 2010).

A less complicated explanation of the impact of forested land on freshwater supply is that such land increases the probability of rainfall. Sheil and Murdiyarso (Sheil and Murdyarso 2009) capture the essence of this relationship in their article under the following revelatory caption: 'how forests attract rain." They contend that 'forested regions generate large-scale flows in atmospheric water vapor' (p. 341). In advancing this empirically-grounded argument, they assign a critical role to evaporation and condensation, which generate atmospheric pressure differences. The impact of forested land on freshwater supply can also be appreciated by understanding the effect of deforestation on rainfall. A number of empirical works have implicated deforestation as a leading cause of the sharp decline in rainfall that has been recorded in parts of Africa (e.g., the Sahel and central Africa regions), South America (e.g., the Amazon) and Asia (e.g., India) (Sheil and Murdyarso 2009).

Western science does not have monopoly over knowledge of the fact that forests possess freshwater making and preservation capabilities. In fact indigenous knowledge systems in Africa and other non-Western regions have recognized this capability for millennia. In Kenya, for instance, indigenous communities around the Mau Forest region in the western part of the country fondly referred to the forest as a 'water tower' (Schwartz 2013). This appellation was not arbitrary; rather, it was inspired by the fact that the forest constituted the main source of freshwater for the Rift Valley and Lake Nyassa (Victoria) region. However, during the last decade or so, the forest has lost as much as 200,000 hectares to agriculture (Schwartz 2013). The result has, among other things, been a significant drop in rainfall and precipitation in the area. Today, the forest has significantly dwindled in size and is a 'water tower' only in name. In other words, the forest is no longer able to fully function as a 'biotic pump.' The theory of forests as 'biotic pumps' is the brainchild of the aforementioned Russian physicists, Makarieva and Gorshkov (e.g., 2007, 2010). The theory essentially contends that forested areas cause a high rate of transpiration resulting in moist air as it ascends. This in turn results in water vapor condensation, which creates a partial vacuum that produces an air pressure gradient. This enables the forest canopy to suck moist air from the ocean. An important aspect of the biotic pump theory is the notion that forest do not simply grow in wet areas; they actually create and nourish the conditions necessary for their growth and sustenance (cf., Schwartz 2013).

Source: The author.

in the region have access to electricity (UN-Energy/Africa 2015: 68). In Sub-Saharan Africa as a whole, only 25% of the population has access to electricity (Brew-Hammond 2010: 2293). However, it is worth noting that there are significant variations with respect to levels of access across the region (Brew-Hammond 2010). On the low end, only 5% of the population in Chad, Somalia, Uganda, Sierra Leone, and Rwanda have access to electricity. On the high end are countries such as Mauritania, Ghana, and South Africa, where more than half of the population has access to electricity. Only one country in the sub-region, namely Mauritius, has an electricity access rate of 100%. At any rate, what is commonplace across the region is the fact that electricity remains more an urban, than a rural phenomenon. A possible explanation for this has to do with the fact that urban income levels are significantly higher than their rural equivalents. It is therefore no wonder that the transition from biomass to electricity, which is driven by income gains (Davis 1998) is faster in urban than in rural areas. Another explanation relates to the fact that the region's electricity grids are restricted to urban areas. The proclivity for concentrating development projects in urban centers is a legacy of colonialism, which equated urbanization with modernity.

Many empirical studies suggest that colonialism explains a significant proportion of the variabilities in the development profiles, including access to basic utilities such as electricity in Africa (see e.g., Njoh 2013; Njoh and Akiwumi 2011; Lee and Schultz 2009; Rodney 1982). These and similar studies tie colonial era policies and institutions to post-colonial differentials in economic growth, public goods/services provisioning, as well as other development indicators (Lee and Schultz 2009). It is conceivable that these variations resulted from the varying investment policies and governance strategies that were employed by different colonial powers. For instance, the indirect rule colonial administrative governance structure is typically associated with British colonialism. This structure is supposed to have endowed erstwhile British colonies with a sense of self-reliance and independence that is rare in former French colonies. On their part, French colonial authorities bequeathed to their erstwhile colonies a centralized administrative structure believed to contain the ingredients necessary for national development.

Despite their shoe-string budgets, colonial governments did everything to make life comfortable for their officials in what they disparagingly called 'the dark continent' (Njoh 2015). By the formal onset of the colonial era in 1884/85, many advances had already been made in the electric power sector. In 1884, at about the same time as the formal onset of the colonial era in Africa, Nikola Tesla invented the electric alternator (AC) (Historical Archive, Online). Tesla's invention made moving electrical current over long distances possible. Half a decade earlier in 1878, an Englishman, Joseph Swan had invented the first incandescent lightbulb. European colonial governments wasted no time to introduce in Africa what at the time were emerging technologies. For instance, British colonial authorities introduced electric power stations in Nigeria in 1886. This is when two electric power generators were set up to serve the Lagos Colony (Sambo et al., Online). In 1951, the colonial government, through an Act of Parliament, created the Electricity Corporation of Nigeria, the country's first formal body in charge of electricity. In Ghana, the British colonial government established the country's first public electricity supply system as far back as 1914 (IDE-JETRO, Online). Initially created in, and to serve, Sekondi, the service was extended to Takoradi in 1928. In Uganda, plans to establish an electricity delivery agency date back to 1905 (Mawajje et al. 2013). This is when the British imperial government had entertained the idea of tapping hydro-electrical energy from River Nile. The idea was later abandoned for lack of effective demand in the country. However, it took more than 30 years for the idea to be seriously reconsidered. Accordingly, in 1936, the Kenya-based East African Power and Lighting Company (EAPLC) secured the license to generate and distribute electricity in Uganda. By 1938, EAPLC had installed one diesel generation station each in, and from which it supplied electricity to, two Ugandan towns, namely Kampala and Jinja (Mwaura 2012). In 1947, a colonial government ordinance, the Electricity Ordinance of 1947 created the Uganda Electricity Board, which acquired the assets of EAPLC and began operating in 1948 (Mwaura 2012: 73). Colonial powers remained active in the electric energy sector in Africa subsequent to the demise of colonialism in most of the continent in the 1960s. For example, the Portuguese, collaborated with a number of German, British and South African companies to build a hydro power plant in Mozambique, a Portuguese colony until 1975. Known as the Cahora Bassa hydroelectric power plant in Western Mozambique, the plant was built between 1969 and 1974 (Encyclopedia Britanica, Online, para. 2).

The introduction in colonial Africa of electricity, an example of cutting edge technology at that time, had two main objectives. One was to create replicas of European towns in Africa (Njoh 2007; Home 1990, 1997). This was intended to attract and retain European colonial officials in the continent. The other was to facilitate colonial development. A definite link was assumed between infrastructure and colonial development. Thus, colonial authorities were patently cognizant of the importance of infrastructure for colonial development aspirations. Roads and railways were necessary to facilitate evacuation of natural resources for onward transmission to colonial master nations at the least cost (Njoh 1999). Similarly, electric energy was necessary to control the temperature in warehouses, hospitals, schools and other facilities that were required to operate the colonial machinery.

Access to Electricity in Africa. Electrical Energy Although there is evidence suggesting that Africans had explored some sources of energy before the European conquest (Mavhunga and Trischier 2014), the history of modern electricity on the continent is inextricably tied to European colonialism (Showers 2011). Initial efforts in this regard were aimed at providing the electrical power that was necessary to operate mines and light European enclaves of the colonial era. Consequently, the earliest modern energy projects were gas- and/or firewood-powered generators that were installed at/or near mining sites and exclusive European enclaves. As far back as the late-1800s, mines and White enclaves in British colonial southern Africa already boasted modern electricity from gas-powered generators. The case of White enclaves in Kimberly and Cape Town in South Africa, and Harare, Southern Rhodesia (present-day Zimbabwe), which were already equipped with electric street lights is illustrative (Showers 2011). This foray into the history of modern electricity in Africa is not only of historical importance; it is also of contemporary significance. As Njoh (2016: 214) has observed, colonial legacy explains most of the "inter-country differentials in electricity supply capabilities of African countries." A number of other researchers (e.g. Lee and Schultz 2009) have suggested that the profiles of countries in contemporary Africa are a function of their varying colonial experiences.

Contemporary Africa is characterized by an acute problem of electricity famine. In this regard, the continent boasts the lowest level of electricity access in the world (Njoh et al. 2018). This problem is especially severe in the sub-Saharan African region. According to some analysts, this region, excluding South Africa, consumes only 68 gigawatts of electricity per annum, with only 25% of the population boasting access to electricity (Njoh et al. 2019: 17). Most households depend on biomass for cooking and heating. In West Africa for instance, biomass accounts for as much as 90% of the energy consumption; only a very small proportion (12%) of the population in this region has access to electricity (Njoh et al. 2019: 17; UN-Energy/Africa 2015: 68). However, as Table 3.1 shows, the energy deprivation problem is not uniform across Africa; the north African littoral region, and the island nations of Seychelles and Mauritius have electricity access rates that are either 100% or close (Njoh et al. 2019: 17).

The problem of low electricity access rate is not inconsequential. Rather, it is at the root of Africa's endemic socioeconomic underdevelopment condition. Some have argued that many countries on the continent would be a lot more developed were it not for their limited access to electricity. For example, Nigeria, the continent's most populous country and largest economy, with a GDP of some 568.5 billion US dollars would be doing a lot better were it not for its poor electricity access record (World Bank, Online). To appreciate the importance of energy in the built environment in SSA region, one must understand it in its comprehensive form as opposed to simply a source of power. Thus, energy in built space must be taken to include all activities necessary for tapping energy from different sources, processing and delivering it to end users. It also includes, in the case of electrical energy for instance, the manufacturing, transportation, repairs, delivering and operation of electrical equipment, as well as the building and operation of electrical facilities. Therefore, no discussion of the importance of energy in built space in SSA can be deemed complete unless it acknowledges the direct and indirect contributions of these activities to socio-economic development. Therefore, the conversation about access to energy in built space in Africa must be broadened to include access to employment and other opportunities generated through the production and consumption energy. From this perspective, it is easy to see how limited access to energy has not only hampered development in SSA; it has also aggravated gender-based disparities in the region (Njoh et al. 2018; Winther et al. 2018). A preponderance of empirical evidence has demonstrated the presence of an inverse relationship between access to electricity and gender-based inequality in educational attainment (e.g., Njoh et al. 2018). This suggests that increased access can result in closing the gap between males and females in many domains. Thus, efforts to increase access to electricity constitute a potentially viable means of combatting prevalent gender-based inequalities in the built environment in sub-Saharan Africa (SSA). The link posited here is easy to understand once one appreciates the fact that women and girls are responsible for most household chores in the retgion. The time that women and girls spend on household chores in the region ranges from one hour per day in Zimbabwe to as many as 5 h per day in Sierra Leone (Njoh et al. 2019: 546). Therefore, any effort to increase access to energy can free up time for women ad girls to indulge in more rewarding activities. This is especially true when access to energy results in the use of electrically powered household appliances such as laundry machines, pressing irons and electric stoves. Although generally taken for granted in the developed world, electric lights are not available in many households in SSA. This makes it difficult for pupils and students to undertake night time studies or complete homework assignments. In some of the major cities where a few street lights are available, students resort to using these for night time studies. The following caption for an article in a major newspaper in the region, illustrates this phenomenon: "Liberia: Where children study under street lights" (Daily Nation 2007). Limited access to electricity also portends problems beyond hampering educational performance; it has far-reaching implications for safety and security in built space. The edited volume on 'sexual abuse of young children in southern Africa' by Linda M. Richter and colleagues contains accounts on inadequate lighting render children vulnerable to assault, rape and other forms of violence in public spaces (see Dawes et al. 2005). Yet, children are not the only real or potential beneficiaries of lighting in built space in Africa. Street and road users in general stand to benefit from such lighting in a region where walking is the primary modde of transportation. It is worth noting that the absence of lights in public spaces has been linked to rising crime rates in many cities in SSA (Gillard et al. 2019). Muncipal authorities and civic groups in the region are increasingly cognizant of the problems that lack of lighting in public and other built space can engender. However, their efforts to adequately respond to the situation are constrained by the lack of financial resources. In fact, street lights in two Ugandan cities, Kampala and Jinja, were recently disconnected because their respective municipal governments, the Kampala Capital City Authority (KCCA), and the Jinja Municipal Council (JMC) seriously defaulted on electricity utility payments (Gillard et al. 2019). Among the many negative upshots of the absence of street lights are an increase crime rates, traffic accidents and night time pedestrian injuries. In addition, the cities have suffered significant economic losses as businesses have limited their activities to daylight hours due to security concerns. Initiatives designed to restore the lights have included enlisting the citizen participation. In this regard, municipal authorities have encouraged local home owners to install security lighting on their properties. In addition, and especially noteworthy for the purpose of the present discussion, both municipal governments have opted in favour of installing solar street lights as opposed conventional ones. This decision, projections show, will result in huge savings for the municipal governments. "Across the two cities, the average cost has been around 6 million UGX (about 1600 USD) per solar street light pole, compared to 8 million UGX (about 2150 USD) for conventional street ligh pole" (Gillard et al. 2019: 9). In Esaghem, a rural village in the South-West Region, Cameroon, village authorities adopted a slightly different variant of the community participation (CP) strategy to take supply their village with electricity from the sun (Njoh et al. 2018). Here, CP entailed enlisting the financial and technical support of natives of the village based in the US. With the labor input of members of the local community, the US-based natives of Esaghem were able to supply solar-powered lights to communal sites and homes throughout the village.

3.9 Concluding Remarks

Sub-Saharan Africa is a product of a triple heritage incorporating elements drawn from indigenous African cultures, Islamic doctrine and Western/Judeo-Christian traditions. Thanks to this rich heritage, the sub-region's historical context emerges as a strong determinant of policies and policy outcomes in most domains. This is particularly the case with policies relating to nature in built space. This is because such policies constitutes a subset of urban or town planning rules and regulations inherited from colonial authorities. This chapter has identified the influence of colonialism not only with respect to the historical context, but also in relation to the political, economic and technological contexts of efforts to promote nature in built space. Among other things, the chapter has demonstrated the intensity of colonial influence on urban and peri-urban forestry, and farming, water supply and energy in the built environment in sub-Saharan Africa. In addition, the chapter has shed light on how contemporary initiatives to develop green infrastructure, address food, energy, and water needs of built areas in the region are building on the foundation developed during the colonial era. Thus, the chapter has uncovered evidence bolstering theories to the effect that a good understanding of Africa's history is necessary to appreciate its current and future development profile.

References

Amoako-Atta B (1998) Preservation of sacred groves in Ghana: Esukawkaw forest reserve and its anweam sacred grove. UNESCO Division of Ecological Sciences South-South Co-operation Programme, Paris

Beach M (2013) Urban agriculture increases food security in Africa. Accessed 20 Aug 2019 at: prb.org/urban-agriculture-poor-africa

Black History Studies (Online) 15 things you did not know about the moors of Spain. Accessed 19 Aug 2019 at: https://blackhistorystudies.com

Boutros C (Online) Top challenges facing Africa today. Accessed 4 Aug 2019 at: https://www.globalyoungvoices.com/fast-news-blog/2016/5/5/top-challenges-facing-africa-today

Brew-Hammond A (2010) Energy access in Africa: challenges ahead. Energy Policy 38:2291–2301

Daily Nation (2007) Liberia: where children study under street lights. Daily Nation, July 27, 2007. Accessed 23 Jan 2020 via: https://www.nation.co.ke/lifestyle/1190-1986-gfcgh02/index

Danso G, Kereita B, Afrane Y (2002) Farming Systems in urban agriculture, Accra, Ghana. With special focus on its profitability, Wastewater use and added Malaria Risk. Consultancy report submitted to FAO-Ghana Office via International Management Institute
Davis M (1998) Rural household energy consumption: the effects of access to electricity—evidence from South Africa. Energy Policy 26(3):207–217
Dawes A, Richter L, Higson-Smith C (2005) Confronting the problem. In: Richter L, Dawes A, Higson-Smith C (eds) Sexual abuse of young children in Southern Africa. Human Science Research Council (HSRC), Cape Town, pp 1–18
Dillon RG (1990) Ranking and Resistance: A Precolonial Cameroonian Polity in Regional Perspective. Stanford, California: Stanford University Press
Economist, The (2007) The Dark Continent: Power shortages have become one of the biggest brakes on development. The Economist Magazine, Available online. Accessed 4 Oct 2019 at: http://www.economist.com/node/9660077
Emeagweli G, Dei S (2014) African indigenous knowledge and disciplines. Brill, Leiden
Enzler SM (2006) Environmental effects of warfare: the impact of war on the environment and human health. Available Online. Accessed 4 Aug 2019 at: https://www.lenntech.com/environmental-effects-war.htmhttps://www.lenntech.com/environmental-effects-war.htm
Gillard R, Oates L, Kasaija P, Sudmant,A, Gouldson A (2019) Sustainable urban infrastructure for all: Lessons on solar-powered street lights from Kampala and Jinja, Uganda An online publication of Frontrunners: a series of policy briefs to inform national governments on the economic and social benefits of action for sustainable cities. Accessed 18 Feb 2020 via: https://www.researchgate.net/publication/332395708_Sustainable_urban_infrastructure_for_all_Lessons_on_solar-powered_street_lights_from_Kampala_and_Jinja_Uganda
Global Trends (2019) Paradox of progress: sub-Saharan Africa. Available online. Accessed 4 Aug 2019 at: https://www.dni.gov/indes.php/the-next-five-years/sub-sahara-africa
Gockowski J, Mbazo'o J, Mbah G, Moulende TF (2003) African traditional leafy vegetables and the urban and Peri-urban poor. Food Policy 28(3):221–235
Gray R (1982) Christianity and colonialism in sub-Saharan Africa. J Black Stud 13(1):59–72
Harvey D (2008) The right to the City. New Left Rev 53:23–40
History World (Online) Accessed 2 Aug 2019 from: History of Africa. http://www.historyworld.net/wrldhis/PlainTextHistories.asp?ParagraphID=oquhttp://www.historyworld.net/wrldhis/PlainTextHistories.asp?ParagraphID=oqu
Home R (1990) Town planning and garden cities in the British colonial empire 1910–1940. Plan Perspect 5:23–37
Home R (1997) Of Planting and Planning: The making of British colonial cities. E & FN, London
Hungerford H, Smiley SL (2016) Comparing colonial water provision in British and French Africa. J Hist Geogr 52:74–83
Kanduza AM (2009) 'Let there be light': the struggle for developing electricity supply in Botswana, 1950–1970. Botswana Notes and Records 41
Kuo, L. (2016). A 700-Year Old West African Farming Practice Could be an Answer to Climate Change. Quartz Africa (June 22, 2016) Accessed online 39–46.on 14 Oct 2019 via: https://qz.com/africa/713512/a-700-year-old-west-african-farming-practice-could-be-an-answer-to-climate-change/
Lee A, Schultz KA (2009) Comparing British and French colonial legacies: A discontinuity analysis of Cameroon. Papers discussed at the Working Group on African Political Economy (WGAPE), Stanford University, 11–12 December. Available online. Retrieved Sep. 20, 2019 from: http://www.sscnet.ucla.edu/polisci/wgape/papers/17_Lee.pdf
Lefebvre H (1968) Le Droit à la ville. Anthropos, Paris
Lefebvre H (1974) La production de l'espace. Anthropos, Paris
Lepapa N (2017) The major problems that Africa is facing today, 2019. Posted online, Wed. Apr 5, 2017. Accessed 4 Aug 2019 at: https://www.africanexponent.com/post/8304-poor-governance-corruption-and-insecurity-major-problems-confronting-africa

Makarieva AM, Gorshkov VG (2007) Biotic pump of atmospheric moisture as driver of the hydrological cycle on land. Hydrol Earth Syst Sci 11:1013–1033. https://doi.org/10.5194/hess-11-1013-2007
Makarieva AM, Gorshkov VG (2010) The biotic pump: condensation, atmospheric dynamics and climate. Int J Water 5(4):365–385
Marwah H (2017) Electricity access inequality in sub-Saharan Africa, 1950–2000. Afr Econ Hist 45(2):113–144
Mavhunga CC (2014) Energy, industry, and transport in South-Central Africa's history. In: Mavhunga CC, Trischier H (eds) Energy (and) colonialism, energy (in)dependence: Africa, Europe, Greenland, North America. Federal Ministry of Education and Research, Munich, pp 5–8
Mavhunga CC, Trischier H (eds) (2014) Energy (and) colonialism, energy (in)dependence: Africa, Europe, Greenland, North America. Federal Ministry of Education and Research, Munich, Germany
Mawajje J, Munyambonera E, Bategeka L (2013) Powering ahead: the reform of the electricity sector in Uganda. Energy and Environment Research 3(2):126–138
Mazrui A (2004) Islam and acculturation in East Africa's experience. Lecture given at the National Defense College of Kenya, near Nairobi, on July 27, 2004, and subsequently repeated at Sheikh Khalifa School, Mombasa, on August 2, 2004. Available online. Retrieved, July 27, 2019 from: http://igcs.binghamton.edu/igcs_site/dirton20.htm
Mbatha S, Wilson Z, Buckley C (n.d.) Zulu indigenous practices in water and sanitation: preliminary field research on indigenous practices in water and sanitation. Paper #199, The Ulundi Water Institute of South Africa (WISA)
Mereri M (2002) Private investment in urban agriculture in Nairobi, Kenya. Urban Agriculture Magazine no. 7. August 2002. Economics of Urban Agriculture. Leusden, RUAF
Mitchell D (2003) The right to the City: social justice and the fight for public space. Guilford Press, New York/London
Mwaura FM (2012) Adopting electricity prepayment billing system to reduce nontechnical energy losses in Uganda: lessons from Rwanda. Util Policy 23:72–79
Nilsson D (2017) The unseeing state: how ideals of modernity have undermined innovation in Africa's urban water systems. N.T.M. https://doi.org/10.1007/s00048-017-0160-0
Njoh AJ (1998) The political economy of urban land reforms in post-colonial state. Int Rev Urban Reg Res 22:408–423
Njoh AJ (1999) The state, urban development policy and Society in Cameroon. Cities Int J Public Policy Plann 16(2):111–122
Njoh AJ (2006) Tradition, culture and development in Africa: historical lessons for modern development planning. Ashgate, Hampshire
Njoh AJ (2007) Planning power: Town planning and social control in colonial Africa. University College London (UCL) Press, London/New York
Njoh AJ (2011) Citizen participation and sustainability: lessons from Cameroon. Development 54(3):376–383
Njoh AJ (2013) Equity, fairness and justice implications of land tenure formalization in Cameroon. Int J Urban Reg Res 37(2):750–768. https://doi.org/10.1111/j.1468-2427.2012.01168.x
Njoh AJ (2015) "The right-to-the-city question" and indigenous urban populations in capital cities in Cameroon. J Asian Afr Stud 52(2):188–200. https://doi.org/10.1177/0021909615570954
Njoh AJ (2016) A multivariate analysis of inter-country differentials in electricity supply as a function of colonialism in Africa. Energy 117:214–221
Njoh AJ, Akiwumi F (2011) The impact of colonization on access to improved water and sanitation facilities in African cities. Cities 28(5):452–460
Njoh AJ, Akiwumi FA (2012) The impact of religion on women empowerment as a millennium development goal in Africa. Soc Indic Res 107:1–18
Njoh AJ, Etta S, Essia U, Ngyah-Etchutambe, Enomah LED, Tarke MO, Tabrey HT (2018) Implications of institutional frameworks for renewable energy policy administration: Case

study of the Esaghem, Cameroon community PV solar electrification project *Energy Policy* 128:17–24

Njoh AJ, Etta S, Ngyah-Etchutambe IB, Enomah LED, Tabrey HT, Essia U (2019) Opportunities and challenges to rural renewable energy projects in Africa: Lessons from the Esaghem Village, Cameroon solar electrification project. Renew Energy 131:1013–1021

Olukoju A (2012) Infrastructure development and urban facilities in logos, 1861–2000. French Institute for Research in Africa (IFRA-Nigeria), Lagos, Nigeria

Phillip A (2006) Electricity and water supply in colonial Nigeria: a case study of Makurdi town 1927–1960. Hist Soc Niger 16:48–77

Prussin L (1974) An introduction to indigenous African architecture. J Soc Archit Hist 58(3):424–433

Rigolon A, Browning MHEM, Lee K, Shin S (2018) Access to urban green space in cities of the global south: a systematic literature review. Urban Sci 2:67. 10.33901urbansci2030067. www.mdpi.com/journal/urbansci

Rodney W (1982) How Europe underdeveloped Africa. Howard University Press, Washington, D.C.

Schneidman W, Ireland J, Kayihura R, Haley B, Owzarkowski U, Fontana L, Crowder S, Stiet K, Mashalane K (2019) Top ten issues to watch in Africa in 2019. Posted Online on February 6, 2019. Accessed 4 Aug 2019 at: https://www.covafrica.com/2019/02/top-ten-issues-to-watch-in-africa-in-2019/

Schwartz JD (2013) Sustainability: clearing forests may transform local—and global—climate. Scientific American, March 4, 2013. Accessed 5 Apr 2018 at: https://www.scientificamerican.com/article/clearing-forests-may-transform-local-and-global-climate/

Selase AE, Jiang LF, Worlanyo AS (2015) Land tenure system in the pre-colonial era: Ghana as the Insight. Int J Afr Asian Stud 14. Accessed 10 Aug 2019 at www.iiste.org

Sheil D, Murdyarso D (2009) How forests attract rain: an examination of a new hypothesis. Bioscience 59(4):341–347

Sheridan M (2009) The environmental and social history of African sacred groves: a Tanzanian case study. Afr Stud Rev 52(1):73–96

Showers KB (2011) Electrifying Africa: an environmental history with policy implications. Georgr Ann B Hum Geogr 93(3):193–221

Soja EW (2010) Seeking spatial justice. University of Minnesota Press, Minneapolis

Tegegne A, Tadesse M, Mekashu Y, Yamir A (2000) Market-oriented urban and Peri-urban dairy systems. Urban Agriculture Magazine, No. 2, Urban Livestock, Leusden, RUAF

UN-Energy/Africa (2015) Energy for sustainable development: policy options for Africa. UN-Energy/Africa publication to CSD15. A UN collaboration mechanism and UN sub-cluster on energy in support of NEPAD

Winther T, Ulsrud K, Saini A (2018) Solar powered electricity access: implications for women's empowerment in rural Kenya. Energy Res Soc Sci 44:61–74

World Facts (2019) African countries by population density. Accessed 4 Aug 2019 at: https://www.worldatlas.com/articles/african-countries-by-population-density.html

Chapter 4
Nature in Built Space in the MENA Region

Abstract The Middle-East and North Africa (MENA) region is unique in many ways. For one thing, built space in the region assumes an unconventional meaning. It is not comprised exclusively of concrete buildings and paved streets. Rather, it includes the modern urban centres as well as hamlets of makeshift shelters, ancillary facilities and the foot tracks connecting the hamlets to the communal fields that serve as the food source for livestock. For another thing, the region is characterized by natural features such as abundant oil reserves that constitute a boon to its economy; at the same time, it contains vast arid and desert areas that present enormous challenges to efforts to promote nature in built space. By analyzing the region, this chapter accentuates the need to contextualize discussions of nature in built space.

4.1 Introduction

Although commonly used in international politico-economic discourse, the Middle East and North Africa (MENA) is not a United Nations Environmental Protection (UNEP) region. Within the UNEP world regional classification scheme, part of the Middle East, comprising a dozen countries, falls under West Asia. The countries are grouped into two geopolitical regions, the Arabian Peninsula and the Mashriq (UNEP 2012: 375). The Arabian Peninsula contains Yemen, Bahrain, Kuwait, Oman, Qatar, Saudi Arabia, and the United Arab Emirates (UAE), while Mashriq includes Iraq, Jordan, Lebanon, the Occupied Palestinian Territories (OPT), and Syria. For the purpose of this chapter, I use the more conventional geopolitical configuration that classifies the Middle East and North Africa as one region. The region includes the following 22 countries: Afghanistan, Algeria, Bahrain, Cyprus, Egypt, Iran, Iraq, Israel, Jordan, Kuwait, Lebanon, Libya, Morocco, Oman, Palestine, Qatar, Saudi Arabia, Syria, Tunisia, Turkey, UAE and Yemen.

The region is unique on many fronts, and analyzing it provides an opportunity to accentuate the need to contextualize discussions of nature in built space. The notion of built space itself is one that requires qualification as it is bound to differ by geo-cultural circumstances. For nomadic populations communities such as the pastoralists that are present throughout the region, built space is not comprised of concrete buildings and paved streets. Rather, it comprises the makeshift shelters, ancillary

A. J. Njoh, *Nature in the Built Environment*,
https://doi.org/10.1007/978-3-030-39759-3_4

facilities and the foot tracks connecting their villages to the communal fields that serve as the food source for their livestock. The natural features of the region are also peculiar and engender problems that differ significantly from what prevails elsewhere in form, magnitude and implications. The unique nature and importance of these features are discussed with respect to five specific substantive environmental matters of utmost importance to the built environment. These include urban land, urban forestry, green architecture, water resource management, energy and food. My main aim in this chapter is to analyze the key socio-cultural, politico-economic, historical and techno-ecological drivers of policies and other actions focusing on these environmental matters. I begin by furnishing some background information of relevance to promoting nature in built space in the region.

4.2 Background

Political Context The main structures of governments around the world can be meaningfully discussed under two main categories, democratic and authoritarian. These structures influence environmental policy making and dictate environmental policy input and actions. In this connection, the environmental policy inputs, outputs and outcomes of democratic governments are called democratic environmentalism while those of authoritarian governments are known as authoritarian environmentalism (Gilley 2012; Wells 2007; Doyle and Simpson 2006; Heilbroner 1974). As an environmental decision-making model, authoritarian environmentalism entails the concentration of environmental decision-making powers in the hands of elites and executive institutions. In contrast, democratic environmentalism connotes the devolution of environmental decision-making powers to different levels and agencies of government. Above all, it entails involvement of citizens in the environmental decision-making and implementation process. This preamble is necessary to understand the implications of the political context of MENA countries for efforts to promote natural resource development and management in their built spaces.

The stereotypical view of MENA countries is that they are all oil-rich and led by monarchs. Yet, nothing could be further from the truth. The region comprises both rich and poor countries with governments organized based on different structures and principles. For simplicity sake, the political systems can be divided into three main groups as follows: Parliamentary republics, which include Algeria, Egypt, Israel, Iraq, Lebanon, Libya, Syria, Tunisia, Turkey, and Yemen; Traditional Monarchies, comprising Bahrain, Kuwait, Oman, Qatar, Saudi Arabia, and United Arab Emirates (UAE); and Constitutional Monarchies, which include Jordan and Morocco. Thus, with the exception of the parliamentary republics, most MENA countries are classified as authoritarian regimes wherein power is concentrated in the hands of a single powerful leader (Gasiorowski and Yom 2017). Only two countries in the region, Tunisia and Israel can be considered fully democratic in the conventional sense. The case of Israel deserves a footnote; the exclusion of its Palestinian minority population from the democratic process constitutes a blemish on its identity as a democratic polity. Otherwise, Israel along with Tunisia qualify as the only democratic polities in the MENA region. Each boasts a system of representative democracy and functions as a republic with an elected president/head of state, a prime minister/head of government, a unicameral parliament and a civil law court system. Thus, liberal democracies are the exception but not the rule in the region. One consequence of this is the absence of a strong civil society. This has far-reaching implications for community participation in efforts to develop and preserve nature in built space.

The form of the state or political system has several far-reaching implications for the development and management of natural resources in built space. For instance, it determines the extent to which the state can succeed in efforts to set aside land for natural resource conservation in built space. In theory, such efforts are likely to be more successful in the monarchies (Bahrain, Kuwait, Oman, Qatar, Saudi Arabia, UAE, Jordan and Morocco) than in the parliamentary republics (Algeria, Egypt, Israel, Iraq, Lebanon, Libya, Syria, Tunisia, Turkey, and Yemen). This is because, while the concentration of power does not bode well for democratic freedom, it holds much promise, at least in theory, for urban planning and the promotion of nature in built space.

Environmental policy making in centralized polities, as suggested earlier, has been labeled authoritarian environmentalism (Gilley 2012; Sowers 2007; Doyle and Simpson 2006). As a policy making model, authoritarian environmentalism is both prescriptive and descriptive. On the one hand, it is prescriptive when it provides guidelines on what countries should do to effectively address the environmental challenges they face. On the other hand, it is descriptive when it paints a picture of what authoritarian countries do in their efforts to address their environmental challenges. The most conspicuous features the authoritarian environmental model include the following (Gilley 2012; Wells 2007):

- absence of inhibitions with respect to the exercise of power;
- decrease in individual liberty; and
- concentration of decision-making power in the hands of an autonomous state.

Each of these features has been shown to be of some advantage to environmental policy making. For example, the absence of inhibitions to the exercise of power, as Heilbroner (1974) argued, stifles the genre of free speech that can impede efforts to control population growth. Consider China's 'one-child policy' (1979–2015), which was designed to especially ease environmental and natural resource challenges. This would have been unsuccessful under conditions of free speech and/or with inhibitions to the exercise of power. Similarly, limitations on individual liberty, which can be criticized on many grounds, has been shown to have some advantages for environmental policy making. For instance, Beeson (2010) contends that such limitations have proved useful in efforts to prevent individuals from indulging in anti-natural conservation behavior. The concentration of power reduces friction and cost in the environmental policy making process.

Many studies have presented empirical evidence demonstrating the effectiveness of these features in efforts to attain environmental policy objectives in some MENA countries. For instance, a study by Sowers (2007) attributes the effectiveness of nature reserves in Egypt to the concentration of state authority in executive institutions. On its part, a study by Doyle and Simpson (2006) uncovered evidence suggesting that success in mobilizing societal support for state environmental policies is a function of the country's highly centralized governance structure.

Ecological and Environmental Context The region is geographically, environmentally and ecologically diverse in terms of elevation, flora and fauna. It includes points as low as 300 meters below sea level (in the Dead Sea region), and as high as 3000 metres above sea level in the mountainous areas in the north (WCPA 2019). Mainly semi-arid, the region faces some of the most challenging environmental problems with the most serious of these being the scarcity of rainfall. One consequence of this latter is the distressing problem of water scarcity that frequently culminates in assiduous spells of droughts. These are exacerbated by increasing demand occasioned by rapid population growth, rising temperatures and decreased precipitation Thus, water emerges as the region's most valuable resource (UNEP 2012).

The region's water problem is vividly characterized by a recent article commissioned by The Water Project, a US-based international Non-Profit Organization (NGO) (Barton 2019). The problem is so severe to the point of requiring immediate international attention in countries such as Yemen, the UAE, Saudi Arabia, Iraq and Jordan. The situation is aggravated for countries such as Jordan and Yemen by their geographical location. The geographical locations of these countries—Jordan in the Syrian Desert, and Yemen at the southernmost end of the Arabian Peninsula—are precarious; they exacerbate the water shortage problem. For instance, Jordan's freshwater withdrawal is less than 10% of Portugal's despite the fact that the two countries boast the same population size (Barton 2019). Another persistent problem with far-reaching implications for the region's water supply in general and efforts to promote nature in built space in particular is desertification. This problem has attained devastating levels in some of the countries such as Iraq, Jordan, Iran and Syria.

Social and Demographic Context The Middle East and North Africa (MENA) region has undergone many significant social changes with far-reaching implications for the natural environment and built space in the recent decades. The most notable of these include increases in the following domains (World Bank 2019; DOC 2019):

- Rates of urbanization;
- Youth (persons 30 or under) population;
- Enrollment of school-aged children; and
- Proportion of population with a college degree.

Urbanization is one of the social changes with the most direct impact on the natural environment in general, and nature in built space in particular, in MENA. Before delving into these impacts, it is necessary to gain some appreciation of the state of the region's urbanization. Table 4.1 contains recent urbanization statistics for

Table 4.1 Levels of urbanization in the MENA region for 2018

Item	Country	%Urbanized
1.	Kuwait	100
2.	Qatar	99
3.	Israel	92
4.	Jordan	91
5.	Bahrain	89
6.	Lebanon	89
7.	United Arab Emirates (UAE)	87
8.	Saudi Arabia	84
9.	Oman	85
10.	Libya	80
11.	Iran	75
12.	Algeria	73
13.	Iraq	70
14.	Tunisia	69
15.	Morocco	62
16.	Syria	53
17.	Egypt	43
18.	Yemen	37

Source: Compiled from UNDESA (2019)

MENA countries. As the table shows, only two, Egypt and Yemen, have urbanization levels below 50%. Thus, a majority of the people in almost all the countries in the region live in urban areas. In some cases, such as Kuwait, the level of urbanization is 100%. Qatar, Israel and Jordan are more than 90% urbanized. These high levels of urbanization accentuate the importance of promoting natural resource development, preservation and management in the region. This is especially true for two main reasons. First, urbanization entails both increases in population and spatial or physical expansion. Second, urbanization places significant demand on scarce natural resources such as land, food, water and energy. A number of recent studies have marshaled evidence to demonstrate the veracity of this assertion for the MENA region in particular. Two examples come to mind. The first is by Topcu and Girgin (2016); this presented empirical evidence suggesting that urbanization leads to a heightened demand for energy. The second, which also interrogates the impact of urbanization on energy is by Al-mulali and colleagues. It revealed that urbanization and energy are reciprocally linked. That is, increases in levels of urbanization lead to increased demand for energy; in turn, higher levels of energy consumption also results in increased levels of urbanization. Furthermore, they revealed that increases in both levels of urbanization and energy consumption jointly cause a rise in CO_2 emission. These findings heighten the need for especially food-producing trees in built space in the region. Such trees can serve three critical functions: they can contribute to purifying the air thereby curbing CO_2 emissions; they can provide badly needed shade given the region's arid conditions; and above all, they can serve as a source of food.

Another demographic trend worth noting is the surge in MENA's youth population. The United Nations Children's Fund (UNICEF) estimates that children and the youth (i.e., persons between 0 and 24) account for about half of the region's population (UNICEF 2019a, b). As for the youth—that is, persons aged 15 to 29—alone, they make up as much as 28% or 108 million of the population (Youthpolicy.org 2019). This is the largest youth population the region has ever known. Moreover, young people, especially the age-group, 15 to 29 is the fastest growing, and comprise as much as 20% of the population in most of the countries, including Algeria, Egypt, Iraq, Jordan, Lebanon, Libya, Morocco, Oman, Saudi Arabia, Syria, Tunisia, Yemen. These statistics are magnified in the Arab countries, where the youth population is the fastest growing. Here, as much as 60% of the population is under 25, while the median age is 22 compared to the global average, which is 28 (Youthpolicy.org 2019).

This demographic trend has far-reaching implications for participatory efforts to create and manage natural resources in the region's built space. These implications may be positive or negative. On the negative side, young people in the region are less likely to prioritize community development projects. Two factors account for this. First, the youth, albeit strong adherents to Islam, do not have the same commitment to Islamic values as the older generation (Gertel and Hexel 2018). Consequently, they may be less likely to commit to major Islamic creeds such as those that entreat Muslims to be actively involved in the affairs of their communities. Second, the growing youth bulge has resulted in rapidly rising levels of youth unemployment, which has led to discontent and social unrest; these can conspire to hinder youth participation in urban natural resource management.

On the positive side, the youth bulge in MENA has three possible effects on efforts to create and manage nature in built space in the region. First, the youth population constitutes a reservoir of physical energy and talent for urban greening projects such as tree planting, parks and green space development and upkeep. Second, the growing youth population is likely to ensure that recreational and other facilities designed to promote nature in built space are used to maximum. The effective use of these facilities as envisaged here stands to rationalize investments in their development and maintenance. Finally, the youth bulge can incentivize authorities to invest in recreational facilities such as sporting arenas and parks. Some countries in the region are already taking such steps. One trailblazer in this regard is Jordan. Here, and particularly in Amman, authorities made a conscious effort to incorporate an abundance of green and pedestrian-only spaces, green areas and youth recreational facilities in the "Greater Amman Master Plan" (Fahmy and Dhillon 2008). With this master plan, which won the 2007 World Leadership Award for Town Planning, the authorities in Amman demonstrated a knack and interest in investing not only in 'visible infrastructure' but also in 'invisible infrastructure.' In many ways, the Greater Amman Master Plan exceedingly succeeds at addressing some important needs of the city's bulging youth population and promoting nature in the city. Fahmy and Dhillon (2008: para. 3) make this point more succinctly in the following words.

> In developing green spaces, local governments are not only investing in physical infrastructure but also engendering a greater social and environmental awareness among their young citizens. In expanding recreational spaces for youth, city governments are providing and promoting creativity and life-skills as the foundation for human capital development.

Yet another social change that MENA countries are experiencing is an unprecedented increase in school-aged children. This is a reflection of the surge in the percentage of children aged fourteen and under; they currently comprise 32% of the region's population (UNICEF 2019a, b, para. 3). The foundation for this surge was laid in the 1970s. In this regard, school enrollment and years of schooling increased in the region by 65% in Arab countries between 1970 and 2010 (DOC 2019). The obvious needs generated by this increase are school facilities, including classrooms and commensurate supplies. Less obvious, but equally important, are facilities that directly impact the inventory of natural resources in built space such as children's parks and playgrounds. The provisioning of these facilities requires significant public expenditure. Evidence suggests that efforts in this connection by governments in the region have left much to be desired. For instance, although children (0–14) make up 32% of the population in the region, only 0.8% of its GDP is spent on efforts to address children's needs (UNICEF 2019a, b, para., 3). Equally paltry are public expenditures on social sectors with a direct impact on children's health in particular and welfare in general.

The United Nations Children's Fund (UNICEF) has drawn attention to the tendency to neglect children's welfare on the part of governments in the MENA region. It proposes solutions in three main areas, child poverty, social protection and public finance for children (UNICEF 2019b). Woefully missing from this list is children's health. This is an area in which nature in built space has an important role to play. There is a preponderance of evidence demonstrating that providing the opportunity and facilities for children to play outdoors can go a good way in promoting good child health (see e.g., Bento and Dias 2017; WHO 2019; Ginsburg 2007). One of these studies is by Bento and Dias (2017). It begins by noting that the amount of time children spend in outdoor play has been diminishing thereby contributing to a more sedentary lifestyle. It presents evidence suggesting that outdoor play promotes cognitive, physical, social and emotional well-being among children (Bento and Dias 2017). Similarly, a study by the World Health Organization (WHO 2019) revealed that children indulging in outdoor recreational activities are less likely to suffer from obesity and related problems. Thus, authorities in MENA would do well to invest in efforts to develop outdoor children's recreational facilities, especially parks, as a means of promoting children's health. Such investments would also go a good way to increase the stock of natural resources, particularly green infrastructure, in built space in the region.

An important social trend also worth noting in the MENA is the ever growing desire for higher education. This, according to one analyst, is a function of two other contemporary trends in the region, namely a population that is heavily skewed towards youth, and a growing middle-class affluence (Kjerfve 2014). Governments throughout the region are investing heavily in education, a sector that consumes almost 4% (3.8%) of the region's GDP (Kjerfve 2014: para. 3). This expenditure appears to be demand-driven as the enrollment rate in higher education throughout the area has also risen. The increase has been most dramatic among women, whose enrollment figures now surpass those of men in many countries in the region, including the United Arab Emirates, Palestine, Libya, Tunisia and Kuwait.

These trends have many positive implications for initiatives to promote nature in built space in the region. At least four of these implications readily come to mind. First, higher education invariably exposes people to the world at large. Thus, graduates have the opportunity of learning and adapting to their specific context, the steps that have been taken to promote nature in built space elsewhere. Second, higher education stands to also endow people with knowledge on the health and other advantages of nature in built space thereby encouraging support for urban greening projects. Third, higher education, because of its liberalizing proclivities, can facilitate participatory efforts to promote nature in built space. Finally, local higher educational institutions can endow graduates with the skills necessary to craft urban development plans that can address the region's most urgent needs. One of these is the need for nature in built space.

Economic Context Although the stereotypical view is that MENA economies are all oil-producing/exporting, they are actually diverse. The economies range widely to include oil-importers, such as Egypt, and oil-exporters such as Saudi Arabia, Kuwait and the UAE. Paradoxically, it also includes poor economies such as Syria and Yemen. The region as a whole registered an economic growth rate of 1.7% in 2018, a rate that is expected to increase to 1.9% in 2019 (World Bank 2019). These growth rates would have been a lot higher were it not for the armed struggles that frequently erupt in the region. Countries such as Iraq and Syria are either experiencing or have recently experienced devastating armed conflicts. On its part, Iran's economic growth has been significantly hampered by U.S.-imposed economic sanctions. Finally, according to some projections, lower oil prices could have serious negative effects on the region's economies (World Bank 2019). These developments have implications for efforts to create and maintain nature in built space in the region.

Technological Context Technology has experienced many rapid changes in all domains especially since the First Industrial Revolution of the eighteenth to nineteenth century. Arguably none of these changes has been as impactful on life in MENA as innovations in water recycling technology. Support to this understandably bold assertion can be found in the old adage that 'water is life.' Water recycling technology is already making a significant contribution to efforts to address the region's water scarcity problem. For example, the Japan-based Hitachi Plant Technologies (HPT), one of the leading water recycling companies in the Middle East has been recycling used water for re-use in Dubai since 2008 (Okuma and Shinoda 2009). In one of its many projects, HPT is collaborating with Dubai's Al Ghurair Group to collect human sewage from labor camps, treating and re-selling it as recycled water. The water is recycled and grouped under two distinct categories, namely medium quality and high quality. The medium quality recycled water is used for irrigation and toilet flushing. The high quality version is used as regular water supply for industrial purposes such as concrete mixing, and supplementary water for district cooling.

The use of such recycled water for domestic purposes in particular, and in built space in general, has met with serious cultural resistance in the MENA region. Here, it is necessary to state that Muslim culture, as informed by Islamic doctrine, prioritizes cleanliness and purity. Consequently, efforts to promote the re-use of used water has encountered some resistance in the region. This is certainly a cultural problem whose severity cannot be overstated. In their reflection on this problem, some experts (e.g., Faraqui et al. 2002) have pondered how authorities can succeed in promoting the re-use of recycled water in a region whose people place a high premium on purity.

4.3 Culture and Environmental Stewardship in MENA

The MENA region has a rich history and culture dating back to antiquity. Unfortunately, this rich history and culture remain largely unknown to many; instead, whatever little is known about the region is replete with inaccuracies. Thus, it helps to undertake a brief comparative analysis of MENA culture vis-à-vis the more dominant Western culture. The main religion of the region is Islam. This constitutes the foundation for most institutions, policies and practices in almost every domain. Several elements of the Islamic doctrine have far-reaching implications for efforts to introduce and maintain nature in built space. Secular environmentalists have criticized Islam as being rather too anthropocentric (Economist 2015). From this vantage point, Islam, like other monotheistic religions is seen as placing human beings as not only separate from, but also superior to, nature. Above all, within the framework of Islamic doctrine, critics contend, human life possesses intrinsic value; in contrast other creatures exist to be exploited by, and for the wellbeing of, humans. There is no shortage of passages in the *Qur'an* and *Hadith* heralding the Almighty God's wonderful creations, and the fact that he placed these under the dominion of humans.

Thus, critics would argue, Islam is not eco-friendly, and can therefore not be depended upon for inspiration to promote nature in built space. Many an Islamic philosopher would fervently beg to differ with the foregoing characterization of Islam's view of nature. Seyyed Hossein Nasr is foremost among the prolific philosophers who have written copiously on this subject (see e.g., Nasr 1968a, b, 1972, 1976, 1989). In fact, Nasr contends that Islam assigns to humans the responsibility of caring for the natural environment. The role of humans in this connection as Nasr contends is that of trustee *(khalifa)*—a role that is stipulated in the Qur'an (2:30). In summarizing this viewpoint, Hope and Young (2019, para. 1) stated that Nasr "sees at the center of Islam a charge to protect the natural world—a world that reflects the higher reality of the transcendent God." Thus, it is safe to conclude that Islamic doctrine not only recognizes the importance of nature; it actually implores its adherents, Muslims, to actively engage in protecting it. In this regard, Muslims are beseeched to respond to environmental threats as part of Islam (Brockopp 2012). The Qur'an instructs humans to (Hope and Young: para. 4): "'Do no mischief on the earth after it hath been set in order, but call on him with fear and longing in your hearts: for the Mercy of God is always near to those who do good' (Q.7:56)." Therefore, contrary to the views of critics, Islamic doctrine is of great utility to efforts to create and manage nature in built space. Islam is also vocal on the role of people as members of the communities they live in the process of creating, maintaining and managing the natural environment in built space. Here, the concept of built space should not be taken literally, thereby limiting it exclusively to space containing buildings, paved roads/streets and commensurate infrastructure. Rather, built space includes any area inhabited by humans as well as the areas upon which they depend for food and other life-supporting supplies.

Yet, Islam is not the only element of culture in the Middle East and North Africa that can influence people's view or receptivity of the idea of natural resource preservation in built space. Some elements of pre-Islamic culture in the region are equally influential in this regard. In fact, many of these elements have been extolled for their promise to contribute positively to efforts to create and manage natural resources in built space. One example of this is the *hema* system. An Arabic word for protected area, a *hema* (*ahmia* in plural), is a system for conserving and managing communal land. There is no maximum size set for land that can be preserved as a *hema*. *Ahmia* measuring in the hundreds of kilometers have been recorded. For instance, one of the prominent *ahmia* in Saudi Arabia, Hema al-Ra'bza, near Dayi'ya, a pasture preserve that was once protected by 'Urmar ibnal-Khattab, in the second caliph was approximately 250 km long (Draz 2019). It was said to contain as many as 40,000 animals, mainly horses and camels grazing at any given moment during the time of 'Uthman, the third caliph (Draz 2019). There are very strict rules governing the use of *ahmia*. Under the rule of 'Umar ibn 'abd Aziz who presided over one of the most famous caliphs in the seventh century, it was forbidden to cut even a single tree branch from a *hema*. Transgressors were subjected to severe punishment such as lashing. Box 4.1 contains further information on the *hema* system.

The practice of *hema* in the Arabian Peninsula predates Islam, and dates back more than two millennia (WCPA 2019). It emerged in response to the need to conserve and judiciously utilize scarce and non-substitutable natural resources. In practice, *hema* was employed to protect common pool areas for grazing during drought seasons, or for the protection of large trees used for communal bee-keeping or for providing shed. Dishearteningly, *hema* has since declined in popularity. For example, the *hema* inventory of Saudi Arabia, the seat of the region's dominant religion, declined from 3000 to a paltry 71 between 1969 and 1984 (WCPA 2019).

Islam and CP in Preserving Nature in Built Space Islam advocates community participation (CP) in natural resource management in built space. It does so to the extent that it encourages its adherents to play active roles in all initiatives to improve the welfare of their communities. Its pronouncements on this specific issue are contained in the *hadith* as well as the *Qur'an*. On this score, Islamic doctrine abhors apathy, lack of interest and indifference. Instead, it implores Muslims to play active roles in their communities. Playing such a role is not optional; it is a requirement of life. Within the framework of Islamic teachings, God does not intervene to improve the lot of any community whose members do not actively participate in efforts to do so themselves. According to the holy Qur'an (13:11), "God does not change the condition of a nation unless it changes what is in its heart." Islamic doctrine can further be seen as supportive of the community participation ethos to the extent that this imbibes within its folds selflessness and duty to each other. In fact, some have opined that Friday prayers or Jum'ah, provides a platform that directly promotes the community spirit among Muslims (Imam.org 2017). The congregation of Muslim for these prayers every Friday provides them an ideal opportunity to meet and exchange pleasantries as well as discuss issues of communal relevance on a weekly basis at a minimum. A fundamental belief in Islamic tradition is that Muslims are each other's keeper. The Qur'an stipulates, "Believers are each other's brothers/sisters. Restore peace among your brothers/sisters" (49:10) (Quoted in Imam.org. 2017).

Another aspect of life with implications for efforts to promote nature in built space through community participation in MENA is the worldview of the region's population. The US Army's intelligence unit has prepared a manual that juxtaposes this worldview with that of Westerners. The worldview reposes on six basic concepts, namely atomism, faith, indigenous versus received tradition, justice and equality, paranoia, and the importance of family over self. With respect to atomism, while Arabs are said to have the tendency to compartmentalize incidents and events, Westerners tend to see these as interconnected. This comparison is overly simplistic, and in fact, misleading; it literally characterizes societies in the MENA region as atomistic in contrast to the 'non-atomistic' variants of the Western world. To appreciate this line of argument it is necessary to understand the concept of atomism. This has to do with the extent of social integration in society. Atomistic societies are characterized by little to no social integration or centralization; they are also weak, and lack political authority and mechanism for reinforcing group social solidarity (Scherer 1982; Barnouw 1961). Furthermore, a unit within such societies can easily sever ties with the larger entity of which it is a part. With the exception of the rapidly decreasing Bedouin and other nomadic groups in the region, MENA contains no societies with these characteristics. Here, it is necessary to state that these groups mainly inhabit areas whose ecological conditions favour population dispersal. This notwithstanding, it is misleading to claim that such groups are incapable of participating in community development initiatives such as the creation and management of natural resources in built space. A few environmentalists have hastened to challenge attempts to portray atomism as antithetical to community building (see e.g., Uyl and Rasmussen 2006). Thus, it is safe to say that the MENA region contains no features that are incompatible with community participation in the creation and management of natural resources in built space.

Another dimension on which most Arabs differ from Westerners is with respect to faith. Arabs, in contrast to Westerners, believe that everything in life is controlled by God's will. Thus, their faith is stronger than that of Westerners. This aspect of the identity of MENA society holds much promise for participatory initiatives to develop and manage natural resources in built space. In this regard, the initiatives can be developed and undertaken by religious as opposed to secular institutions. As stated above, the Qur'an contains many passages that place the charge of environmental stewardship squarely on the shoulders of Muslims. In this regard, Islamic doctrine views creatures on Earth and the cosmos as a mark of God's omnipresence. Therefore, taking care of these creatures is one way by which humans must worship God. This is an aspect of the Islamic worldview that is often ignored. This worldview implores Muslims to throw their full weight behind sustainable development initiatives (UN Environment 2018). These initiatives include, *inter alia*, greening built space.

With respect to received, as opposed to indigenous, tradition Arabs are very suspicious of received, particularly Western tradition. In fact, they view Western tradition and culture with skepticism. This problem is a function of centuries of Western efforts to dominate MENA followed by many decades of armed confrontations often led by the United States of America. This may also explain the reason why people in MENA are labelled as paranoia with respect to strangers, especially Westerners. A careful read of this reveals that the issue is not with the message. Instead, it has to do with the messenger. Thus, it is not simply a desire to adhere to Islamic traditions. This suggests that efforts to promote natural resource development and management in built space are likely to succeed when they are led by natives of the region than otherwise.

When it comes to questions of justice and equality, most people in the MENA region place much value on these especially in relation to fellow Muslims. This feature of the identity of the population of MENA can prove exceedingly useful in efforts to involve MENA citizens in urban natural resource management initiatives. This is especially true for one variant of natural resources, namely parks. Parks are a scarce and precious commodity in urban areas, especially in the arid space of MENA. Thus, they are often located in the wealthier neighborhoods. This constitutes a glaring example of environmental injustice especially because parks and other green spaces can improve the wellbeing of urban residents (Silva et al. 2018). Efforts to reverse the situation in the region are likely to succeed if they invoke Islamic doctrine calling for justice and equity among Muslims.

As for the place of family in relation to one's self, communities in the MENA region, emphasize the family over the individual. In particular, families and communities in this region are tight-knit. Thus, the honor and the pride of the family or community is prioritized over that of the individual. This attribute of the dominant MENA ethos is of enormous utility for efforts to promote natural resource development and preservation in built space. MENA communities may range from small hamlets to large urban neighborhoods. A common denominator in these communities is the fact that they have designated leaders. A logical starting point for efforts to 'green' built space is to identify these leaders who would in turn serve as the liaison between the communities and the authorities. Such an arrangement is likely to save cost and promote efficiency and effectiveness. As a cautionary note, the concept of leadership as employed here is not confined to those belonging to the formal sector such as government officials. Rather, this includes traditional leaders such as Imams and local village chiefs. Such leaders are endowed with knowledge of the local environment, customs and traditions.

4.4 Land in Built Space

One defining characteristic of land is its scarcity. This feature peaks for usable and especially habitable land, which is in exceedingly short supply in arid and semi-arid regions such as the Middle East and North African countries. Here, areas for nature must compete for space with housing, transportation, commercial and office complexes, as well as industrial parks. Thus, land stands out as the most important natural resource in built space everywhere but especially in the MENA countries. It is, therefore, no wonder that urban land rights rank as one of the most critical issues in the region. These rights have evolved over centuries and have been shaped by indigenous customs, Islamic doctrine, colonial or received principles and practices, and post-colonial state policies. Thus, the urban land policies of the region can be conveniently discussed under three categories, including indigenous customary system, Islamic land laws, colonial/Western laws, and post-colonial land laws (cf., USAID 2016).

Indigenous Customary System This system, which predated Islam, has evolved over more than two millennia. It can be divided into three main categories according to their land acquisition processes. Under the first category, land is acquired by way of the principle of first cultivator. This bears a striking resemblance to the principle of first occupancy, which was a prominent means of land acquisition in pre-colonial Africa. Like the principle of first occupancy, the principle of first cultivator assigns right of private ownership to the first cultivator of any piece of land. The second category comprises land that was acquired through conquest. In this case, a stronger group typically over-ran, conquered and confiscated the land of a weaker one. The third category includes previously unoccupied land upon which a migrating tribe or clan under the leadership of a chief claims and occupies. Lands in the last two categories were commonplace thereby, making group, communal or collective ownership of land the most dominant indigenous land tenure system in MENA.

Accordingly, villages comprised the dominant landholding units in ancient MENA (Wily 2018). However, as of 10,000 BCE, there was a trend towards individual and corporate ownership of land. This trend was fueled by factors such as population growth, urbanization, and commensurate forces as well as increased territoriality. More recent drivers of this trend have included the need to protect prime land, the increasing adoption of Western culture with capitalism and the commodification of land as two of its most portent elements. In the recent past, there have been some signs that communal landholding may be regaining its once lost allure in the region. Some support for this assertion is provided by the following developments (Wily 2018).

- Increasing recognition of communal land as property although it can neither be individually owned nor sold.
- The growing tendency to recognize communal land of both the registered/titled and unregistered/untitled variants.

- A growing tendency to formalize communal land ownership through the issuance of land titles and certificates to landholding collectivities.
- The enactment of laws and other pieces of legislation specifically designed to govern communal lands.

However, despite these trends, it is necessary to note that customary landholding systems are largely confined to rural areas. Thus, their orbit is mainly agricultural lands, which fall into two principal categories (FAO 2019). The first includes land under the control of ethnic, sectarian or tribal groups. These are lands that have typically been acquired through inheritance or marriage. The second includes cultivated lands in communally held areas. Two prominent features of the customary land tenure system in the MENA region are worth noting here (FAO 2019). One is the lack of a clear-cut demarcation between collective and individual rights over any given parcel of land (FAO 2019). Thus, groups and individuals are free to access and use equitable shares of the collectively owned rain-fed land based on prescriptive rights or allocation. The other is the egalitarian ethos, which ensured that everyone, regardless of societal status, has equal access and right to land within the communal landholding framework. Indigenous land management systems of the MENA region have proved to be very resilient. This resilience is manifested in the fact that many features of the system such *hema,* remain in vogue today (see Box 4.1).

Box 4.1 The Hema Land Conservation/Management System

As a system for grazing and land conservation, *hema* dates back to antiquity; it remains in vogue today. Like in the past, *hema* are controlled by individuals, tribes, or the state. In Saudi Arabia, a 1953 decree placed some *ahmia* (i.e., *hema* lands) at the disposal of the public for free grazing; others remained under the control of tribal leaders. A recent assessment of the two types of *hema* lands—the ones under the state or government and those under the charge of indigenous authorities is very telling. It reveals that the *ahmia* controlled by the state were less productive than those under the authority of indigenous leaders. This is indicative of the suitability of indigenous institutions as the rightful authorities to control *hema* lands.

Ahmia can be grouped into six categories according to their use and/or type of protection they afford the natural resource in question. These include, *ahmia* that serve as a:

1. Source of grass but not on-site grazing. The harvesting of grass to feed cattle is permissible from such sites during certain periods, especially during droughts. The power to grant authorization to access such sites belongs to the traditional or tribal leader.
2. Seasonal source of grass and on-site grazing. Two examples of such *ahmia* are *Hema Elazahra* and *Hema Hameed* around Belgurashi in Saudi Arabia.

(continued)

Box 4.1 (continued)

3. Site of year-round grazing and hay or grass harvesting. For grazing, only specific types and number of animals are permitted. Most ahmia of this type can be found around Taif, and the only animals permitted are donkeys. However, there are no restrictions on hay harvesting as long as the grass is mature.
4. Beekeeping reserve. This is an endangered type of *ahmia* as there are only a few in existence today.
5. Tree reserve meant to protect specific species of forest trees such as juniper, acacia, and ghada. *Ahmia* of this genre are typically communal property as they belong to whole villages or tribes. No tree-cutting is allowed except for emergency purposes such as post-disaster reconstruction or the construction of a mosque or school. Occasionally wood from the trees can serve as a source of funds required to address an urgent need of the village or community as a whole.
6. Natural reserve for one or more villages or tribes. The responsibility for managing reserves of this sort falls on the shoulders of the village or tribal chiefs.

Apart from its function in guaranteeing the productivity of land in built space, *hema* occupies an important place in Islam although it is worth noting that this land management system predated Islam. Prophet Muhammad, the father of Islam, is said to have been concerned with fodder reserves, which he believed were necessary for survival of the Islamic nation. He saved a few *ahmia* of his own, one of which is *Hema Alnaquia*. This is a wadi near Medina that was used mainly by animals to defend the cause of Islam. He declared that *hema* is exclusively for God and His Prophet. This was later construed as a message to governors and other highly-placed agents of the state to protect the *ahmia* in the best interests of the community. At the core of this is an Islamic principle that implores Muslims to view themselves as partners in water, fire, and ephemeral range. More noteworthy, are the pronouncements of the Holy Qur'an that have been invoked to buttress the *hema* system. One such pronouncement portrays the Qur'an as not only an advocate of conservation; it also forbids tree-cutting as well as the destruction of crops. More importantly, it forbids the destruction of nature even in war times.

Source: Adapted from Draz (2019)

Islamic Land Law Until the advent of European colonialism in the late-nineteenth century, land and property rights in the Middle East and North Africa region were largely governed by Islamic principles. The earliest and best-known attempt at codifying and applying these principles to land rights did not occur until 1858 (Rae 2002). This is when authorities under the Ottoman Empire began formalizing land rights in the region. The formalization process involved undertaking cadastral surveys, demarcating, mapping and registering land mainly in the settled portions of the region. Thus, historically, Islamic land law has mainly focused on settled areas, and has had little to do with steppe and desert areas (Rae 2002). However, within the framework of this law, most agricultural land belonged to a category known as *miri* or *amiri*, which means 'property of the Amir.' The law divides privately-owned land *(Mulk)* in permanently irrigated areas, orchards, and house-plots into two main categories (Rae 2002). The first is concerned with resource ownership and control *(raqabah)*, while the second has to do with usufructuary rights *(tassaruf)*. The first, *raqabah,* is essentially the province of the state or governing authority, while the second, *tassaruf,* belonged to the farmer. In most cases, the usufructuary or use rights belonged to more than one farmer—typically, a collectivity of farmers who used the land on a communal basis *(musha)* for grazing.

Three other important classes of land under Islamic real property law are, *Matrukah* (i.e., 'given over') land, *Waqf* (religious land), and *mawat* (dead land) (Rae 2002). The *Matrukah* comprises two categories, namely lands set aside for use in developing public infrastructure such as highways, and land for residential and ancillary facilities such as communal forests, herding stations, public parks, and recreational facilities. The Waqf lands under Islamic land law are set aside to be used exclusively for religious purposes. Such lands could also be placed at the disposal of benevolent or philanthropic entities. The *mawat* lands 'dead land' or unusable land such as steppe and desert areas. Based on Islamic doctrine, particularly *Shari'ah* Law, and the Ottoman Land Code, *mawat* lands were 'open access,' that is, they were opened to everyone and were non-taxable. Such lands were placed at the disposal of all citizens to use as a source of fuel wood, herbage, and so on.

Modern Land Tenure The Food and Agricultural Organization (FAO) has produced a report that provides details on MENA's modern land tenure system.

This system is a product of its historical legacy. Thus, the system contains elements rooted in its indigenous or pre-Islamic tradition, Islam, the Ottoman period, the European colonial era, and the current independent period. However, credit for the earliest formal law goes to the Ottoman authorities who enacted the region's first Land Code in 1858. Originally derived from Shari'ah Law, the Code sought to formalize and clarify land rights by requiring the land certificates and titles as a means of validating all claims of entitlement to any piece of land. Later modifications to the Code were designed to accommodate these claims as founded not only in secular, but also non-secular principles. By the start of the First World War, Ottoman authorities had succeeded in formalizing claims of entitlement to land for only a few parcels of land. In fact, the formalization process was discontinued almost everywhere save in the areas that fell under the administrative jurisdiction of colonial authorities.

These were essentially the urban areas and their peripheries. This latter comprised mainly the areas under cultivation by urban dwellers. Land in such areas were cultivated on a communal basis. However, the land ownership system promoted by colonial authorities recognized exclusively individualized land ownership. Thus, an essential achievement of the colonial authorities was the supplanting of the indigenous, with the individualized land ownership regimes. This resulted in the displacement of members the indigenous population who were previously able to access land through their community membership. Instead, land in urban and adjoining areas throughout the region was concentrated in the hands of the local elite, members of the European settler population, and corporate bodies. The land tenure formalization process had the same effects everywhere in the MENA region. For instance, the French employed the process to endow Europeans with control and ownership of urban land in Tunisia and Morocco. This was the same situation in Algeria, where French colonial authorities used the land registration system to confiscate large swaths of fertile and prime land for themselves.

To have some sense of the implications of land tenure regimes on the promotion of nature in built space in MENA, it is necessary to understand the concept of state lands in the region. This is because the amount of urban land directly owned by the state significantly influences the state's ability to control land use in urban space. Most of what constitutes state lands in the region was known as *Mawat* or dead land during the pre-colonial era. The concept of state land in MENA is theoretical due to the fact that governments in the region own all land within their respective territories. To the extent that this can be actualized, it has far-reaching implications for the creation and management of nature in urban space in the region.

4.5 Water Resources

Paradoxically, the region's water scarcity problem cannot be addressed with the infusion of financial resources alone. This is not to discount the importance of such resources. Rather, it is to say that financial resources are necessary but certainly not sufficient to address the problem. Here, it is worth noting that the problem has thus far proved intractable despite the fact that the region contains some of the world's richest countries such as Saudi Arabia, the United Arab Emirate and Kuwait. In line with the central theme of this book, this chapter proposes community participation (CP) as a viable strategy for addressing the region's water problem. The viability of this strategy is accentuated by the fact that many facets of the region's water scarcity problem have their roots in human actions. This assertion is bolstered by the United Arab Emirates' water demand and supply situation as elucidated by Barton (2019). The UAE is well-known for its skyscrapers, luxurious resorts and lavish attractions. The UAE is reported to have earned the dubious reputation of being the highest per capita consumer of water in the world a decade or so ago (Barton 2019). At this rate, the Emirates risk depleting its natural freshwater resources within the foreseeable future. Addressing the problem would require a lot more than the huge financial outlays that have already gone into developing desalination plants. Initiatives to commit financial resources to this problem must be complemented by changes in human behavior—particularly, an adjustment in water use habits. Community participation strategies have proved exceedingly effective in these circumstances. This is exemplified by the effectiveness of water users associations (WUA), which have been active in the agriculture domain in the region (Ghazouani et al. 2012, also see Box 4.2).

Box 4.2 Water Users Association in Egypt

Egypt has two main sources of water, namely the River Nile and groundwater. The country's official water quota from the Nile is 55.5 billion cubic meters per year (55.5 billion m^3/year). The river serves mainly the Nile Valley and the Delta. Groundwater from two main aquifers, namely the Valley and Delta aquifer and the Nubian Sandstone non-renewable aquifer in the Western Desert. These sources are depended upon for many purposes but especially for agricultural irrigation. Farming functions around 2 to 3 cropping seasons per year in the country on two types of arable lands. The first comprises Old Lands, which are located along the Nile Valley and Delta regions, and New Lands, located west and east of the Delta in the Sinai and oases. Throughout history, water delivery down to the level of the branch canal has been the purview of the Egyptian state. Meanwhile, control over management and distribution at the tertiary level canals, that is, the Mesqa, was delegated to farmers. This is effectively a participatory model of water management. In this case, the model takes on many forms, including *munawaba, mtarafs,* and *sagia* ring for collective water pumping of water. These indigenous traditional systems constitute the nucleus around which the state has encouraged the development of private water associations in the country. Particularly, the Ministry of Water Resources and Irrigation (MWRI) has strived to encourage farmers to play critical roles in irrigation management related water distribution activities. The first formal efforts in this connection involved the creation by the MWRI with funding from the United States Agency for International Development (USAID) of a number of Water Users Associations (WUA) in the late-1980s and early-1990s. The specific projects involved were the Egypt Water Use and Management Project (EWUM), Irrigation Management Systems (IMS) project, and the Irrigation Improvement Projects (IIP). The legal framework for the WUAs is Law No. 213 of 1994, which defined WUAs as legal private organizations at the *mesqa* level. There were as many as 600 WUAs operating in Egypt with about 15% of the country's irrigated areas involving some 500,000 farmers and residents in the early to mid-2000s. Law No. 213 of 1994 also introduced Water Users Unions (WUUs). Within this framework, WUAs and WUUs are operated and controlled by their members for their own benefit. The WUAs operate in the Old Lands while the WUUs operate in the New Lands.

WUAs and WUUs have proven to be an effective means of shifting part of the costs of water management onto farmers; it has emerged as a viable tool

(continued)

Box 4.2 (continued)

for dealing with equity issues, and the resolution of conflicts among users in irrigation systems. They have been responsible for the single-point lift pumping stations introduced by the IIP project.

Apart from Americans, the Dutch have also been active in the water sector in Egypt. In particular, the Dutch Cooperation has promoted water boards (WBs) at both the branch canal and district levels.

Source: Ghazouani et al. (2012)

The versatility of WUAs and WUUs notwithstanding, the most important element of a participatory model remains the people themselves. The importance of people is underscored in projects that target water distribution systems in the region. In one such project, it was revealed that users at the head of a canal irrigation district were tapping more than their fair share of water (Faraqui 2001). The users were said to have employed illegal pumps to extract water directly from the system. The participatory strategy to resolve this problem involved the use of publicity campaigns. Part of the campaigns included messages from Imams to mosque attendees during regular Friday prayer meetings. The messages drew on the Qur'an to admonish against taking another person's share of water. This, according to Qur'an, is patently sinful and morally reprehensible. This strategy, as Faraqui (2001: 6–8) recounted, proved to be very effective given that the complaints about water shortages downstream of the water distribution system decreased by 32%. Also, with 1978 *fatwa* in Saudi Arabia, water reuse in the country increased; and in 1995, the country reported reusing 15% of its treated wastewater to irrigate date palms and fodder, such as alfalfa. In addition, water used for religious rituals in mosques in Mecca and Medina is used for toilet flushing. Furthermore, recycled water is used to irrigate more than seventeen hectares of alfafa, garlic, onions, aubergines and peppers in Kuwait. Jordan has also registered encouraging results in its efforts to promote recycled water usage. As many as 70 cubic metres of recycled domestic wastewater were reused for restricted irrigation; this was 12% of all of the water used for irrigation that year.

These revelations attest to the functional role that public awareness campaigns can play in water conservations. However, for such campaigns to work, they must be holistic and multidisciplinary. Thus, they need to target other communities and venues; not exclusively Muslims and mosques. In fact, as the Egyptian National Community Water Conservation Programme (NCWCP) once recommended, "the strategy of water conservation must be global and interactive and include all actors concerned, such as religious, political and informal community leaders" (Faraqui 2001: 7 and Faraqui et al. 2002).

In emphasizing the need for a global awareness education as a strategy for maximizing the utility of community participation in the water domain in MENA, Faraqui et al. (2001) makes the following recommendation. Authorities would do well to:

> Conduct a wide-ranging pilot study to integrate religious elements into a comprehensive programme of public education and awareness projects to encourage conservation and reuse, with particular emphasis on women and girls, who are often left out of such programmes because their religious learning does not occur in mosques or schools (Faraqui 2001: 25).

No narrative on the water question in MENA can be considered complete without a word on used water recycling. As stated earlier, the science and technology for recycling used water is at an advanced stage; they hold much promise for efforts to address MENA's water scarcity problem. However, success in this connection is significantly compromised by serious cultural resistance to the use of recycled water. The use of recycled waste water is considered undesirable, and even characterized as *haraam* or sinful according to Islamic doctrine. Efforts to overcome this resistance to recycled water in the region have included the execution of scientific studies to determine and ascertain the quality, usability and safety of this water. One such study, which was conducted in Saudi Arabia concluded that recycled water is in fact safe and usable. On their own, the scientific findings did not succeed in overcoming people's resistance to recycled water. Consequently, it was necessary to enlist participatory strategies, which included input from religious and community leaders. The public awareness and education messages from these leaders essentially conveyed the following sentiment as a water demand management strategy:

- The need to conserve water is emphasized by Islam;
- The use of thoroughly recycled and treated water is acceptable by Islam;
- Islam recognizes the scarcity of water and preaches against it being wasted.

4.6 Forests and Greening in Built Space

Like many concepts in urban planning and environmental policy, the act of urban forestry and greening has been defined variously by different analysts. According to Akerlund et al. (2006), it "refers to all activities related to the whole urban green resources." For the purpose of the discussion in this book, the act of urban forestry and greening is defined as all activities designed to promote the development, preservation and management of green resources in built space. There are three types of green spaces in the MENA region including the following (Hashem 2013):

- Public parks;
- Private green;
- Natural vegetation.

Public parks, as the name suggests, belong to the state or municipal governments, and are openned to the general public. Private green spaces are owned and controlled by private entities. Consequently, they are not openned to the general public. These are typically located within residential facilities, gated communities, and private farms. Meanwhile natural vegetation comprises sporadic seasonal desert vegetation. This tends to grow mainly in land depressions. It has a short life following the rainy season.

Trees have always been treasured as part of the built environment in the region. They appear in many stories and metaphors in the Holy Bible. For instance, in the New Testament Book of John (12: 13), it is reported that Jesus Christ triumphally entered Jerusalem on Palm Sunday through streets that were decorated with leaves from palms and other trees. The Old Testament Book of Ezekiel (31: 3) states that, "Behold, Assyria was a cedar in Lebanon with beautiful branches and forest shade." Modern historical evidence of the priority place of trees in built space are also abundant. For example, archival data from the seventeenth century reveal that Ottoman Empire authorities promoted the planting of trees that possess spiritual value, and/or those that can produce edible fruits or provide shade and shelter (Akerlund et al. 2006).

More recently, tree planting and greening programs have been widely adopted as a major component of urban planning throughout the region. In some cases, the practice of tree planting and greening in built space is not only connected to history but also to the culture of some towns and cities in the region. The case of Tehran, whose name connotes the place of plane trees in Persian is illustrative. The earliest urban forest plantation in this city was established in 1935; its main purpose was to control dust, reduce air pollution, control erosion and provide some forest products (Akerlund et al. 2006).

In Tehran, a city whose name in Persian actually means 'the place of the plane trees'.

Although the avowed purpose of activities to maintain trees and other natural features in built space in the region has not changed, only the wealthy countries are actively involved in such activities. Witness the fact that Yemen, an impoverished country, has witnessed the problem of severe decline, and in some cases, the complete disappearance of their stock of urban forests. Apart from poverty, the absence of realistic forestry policies and rapid rates of urbanization without corresponding levels of economic development have contributed significantly to this problem. Yet, even the poorest of these countries have made some effort to cultivate trees and green spaces in their built space. In Yemen, for instance, the cities of Aden and Al-Hodeydah are circumscribed by green belts. These are especially noteworthy because of their extensive length; was 14 kilometers and 100 meters was in the early-2000s (FAO 2002).

As for the oil-rich countries of the region, they have not only crafted well thought-out urban afforestation and greening policies, they have also prioritized the implementation of these policies. These initiatives have been undertaken under the rubric of urban beautification and environmental improvement. A quick examination of the experiences of Bahrain, Kuwait, Oman, and the United Arab Emirate in this connection is informative. The built space, especially the cities of these boast carefully planted trees, flowers and fodder that adorn and have transformed roadsides into gardens, and roundabouts into mini-parks (Akerlund et al. 2006; McBride and Mossadegh 2000). Over the years, green areas throughout built space in these countries have improved not only in qualitative terms but also quantitatively. For instance, in 1974, Abu Dhabi, one of UAE's largest cities, boasted only one major park. Today, the city has more than 40 parks covering an area of approximately 300 ha (Akerlund et al. 2006).

It is generally agreed that forests and related green spaces contribute significantly to the aesthetic appeal and wellbeing of built areas. Yet, it is often not clear what constitutes urban forests and greening. A useful operational definition of such spaces would include the following elements (Akerlund et al. 2006):

- Street trees and road plantations;
- Public green areas such as parks, gardens, and cemeteries;
- Semi-private space, including green spaces in residential areas, office and industrial parks;
- Orchards, botanical gardens, and cognate agricultural outfits;
- Forest reserves.

An important determinant of the success of urban forestry and greening initiative is the availability of fertile land. As stated earlier, the MENA region is comprised mainly of arid and semi-arid, which is largely infertile. This poses several challenges to efforts designed to green built space in the region. Barrett and Leal (2017) identified and discussed three of these to include, the lack of space, poor soil conditions, and insufficient and inconsistent rainfall. This does not mean that cultivating forests and green space is impossible in the region. Rather, it means that the cost of doing so is exorbitant. This explains the fact that the oil-rich countries are those that have registered the most progress in this regard in the region. The concept of success here is relative as no country in the region actually meets the World Health Organization (WHO) stipulation of 12 square metres of green space per capita in urban areas (Akerlund et al. 2006). Not only do cities in the region significantly deficient in terms of available forests and greenery in built space, they do not have statistics on the area of such space. Tehran, Iran is one of the few cities that keep records of these statistics. The city was reported to have had about 3 square metres of green space in 2000 (Akerlund et al. 2006).

The space problem is a function of the fact that that streets and sidewalks especially in planned built space have fixed dimensions. This problem is compounded by the presence of utility lines and commensurate fixtures such as utility poles, fire hydrants, utility boxes, parking meters as well as traffic signal poles and lights. These fixtures take up space that would otherwise be ideally suited for street trees. Also, overhead utility lines and poles tend to interfere with urban tree-planting initiatives especially with respect to trees designed to provide overhead canopies. Thus, authorities would do well to make provision for trees at the urban design stage. In built-up areas requiring retrofitting, taking overhead utility cables underground would constitute a workable option; such a strategy would also eliminate the need for utility poles.

As for the soil conditions, arid soils are typically never ideal for trees. In fact, only certain trees survive in arid conditions, and authorities must endeavor to carefully identify such trees. Typically, these would be trees such as acacia, ash, buckeye, catalpa, cedar, crape myrtle, eucalyptus, hackberry, juniper, oak, pine, poplar, and silver linden, which have been identified heat-tolerant (England 2019). The precarious geography of the region has combined with other factors such the failure to prioritize greening by authorities to stifle progress on expanding green space in some countries. In fact, the green/built up area ratio, which is critical for urban quality of life remains low in both the wealthy and poor countries in the region. For instance, a study of Greater Doha in the oil-rich country of Qatar, revealed that the park provision ratio ranged from 0 to 0.03 through 0.43 and 0.48 to 0.92 and 0.99 hectare per thousand (1000) population with an average of 0.37 (Hashem 2013). These, as the study noted, falls significantly below the standard adopted for public parks. Thus, the city is experiencing a problem of urban green space (UGS) scarcity. This is due to rapid rates of urbanization.

4.7 Built Space, Agriculture and Food Security

The relevant literature treats matters relating to agriculture and food security in built space under the general rubric of urban agriculture. This latter refers to farming and related activities that take place within, and in close proximity to, urban areas. Such activities have always pre-occupied people in the MENA region throughout their long and rich history. One factor at the root of agriculture's primordial place in this history is the nature of the region's geography. As mentioned earlier, the region is largely semi-arid and arid. This renders agriculture extremely difficult. Consequently, the region has known more than its own share of food insecurity problems since antiquity. In fact, the Holy Bible, arguably the most authoritative source of the region's written history, contains a plethora of stories on how people in the region faced, and worked successfully to address these problems in ancient times. One such story, summarized in Box 4.3, is about how Joseph drew on his ingenuity to help King Pharaoh address a critical famine problem in ancient Egypt.

Today, more than two millennia since Egypt's devastating famine that was reported in the Old Testament of the Bible, authorities continue to work indefatigably to guarantee food sufficiency for the region. Although the region is a net importer of the food it consumes, a significant proportion of its food is locally produced. The main farming systems responsible for this production include the following (Dixon et al. 2001):

- Irrigated farming;
- Highland mixed farming;
- Rainfed mixed farming;

Box 4.3 Joseph's Agricultural Policy Solved EGYPT'S Famine Problem
The Old Testament book of Genesis (41: 47–58), narrates the story of Joseph's dream of, and successful effort to address, Egypt's famine problem. Egypt was experiencing a season of abundant harvest; however, given the region's precarious geography, Joseph devised a plan to set food aside for moments of poor harvests. The plan consisted of having the king (Pharaoh) fund a program to construct and stock granaries in cities throughout Egypt. The period of abundant harvest ended after 7 years when famine hit the land of Egypt. By this time, the granaries were overflowing with grain. The famine grew so severe that it wreaked havoc throughout the land of Egypt, the region of Goshen where the Israelites had been resettled, Canaan and neighboring territories. To feed the population during the difficult times, Joseph tapped grain and other food from the storage facilities. People were charged a nominal fee, and when they ran out of money, they were allowed to barter their livestock for, food. When the people no longer had any livestock, they began bartering their land, and those without land, offered themselves by literally entering servitude in the service of the King in exchange for food.

Source: Genesis 41 (NIV)

- Dryland mixed farming;
- Pastoral farming;
- Sparse arid farming;
- Coastal artisanal farming;
- Urban-based farming.

Of interest for the purpose of this book is the last system, urban-based farming. Thus, the other systems fall beyond the book's scope. Suffice to highlight the following important pieces of information about the systems (Dixon et al. 2001). Irrigated farming engages 17% of the farming population, and 2% of the agricultural land in the region. It depends largely on surface water from rivers such as the Nile and Euphrates. Highland mixed farming is the most important; it engages as much as 30% of the farming population, and occupies 7% of the agricultural land in the region. Rainfed mixed farming engages 18 cent of the farming population, and occupies 2% of arable land in the region. Dryland mixed farming takes up 4% of the arable land, and employs 14% of the farming population; while pastoral farming occupies 23% of arable land, and employs 9% of the agricultural population. Coastal artisanal farming takes place mainly around the Mediterranean Sea, and occupies only 1% of the arable land while engaging about a percent of the agricultural population.

Urban-based farming, which as noted earlier, has always been crucial throughout the region, continues to play an important role. Although it occupies less than 1% of the arable land, it engages as much as 6% of the agricultural population. Urban-based farming involves persons working in many farming areas, including individually- or family-owned private gardens, private grazers on privately-owned or communally-owned fields, state-owned and operated state-of-the-art gardens, and collective farms or grazing fields.

Individually- and family-owned urban-based farms are the most common in the region. It typically involves individuals or families cultivating vegetable and other types of food plants in open spaces dotted around the urban centres, including the backyard of individual or family homes. Occasionally, the garden occupies areas that were once at the outskirts of a city, but have since been encapsulated within built space. Thus, urban gardens are essentially remnants of agriculture land that have ended up being trapped amongst high rise buildings, villas, and shopping malls (Zurayk 2010). Typically individually owned, such gardens constitute conspicuous features of the landscape of major cities such as Beirut, Lebanon (Tawk et al. n.d.) and Amman, Jordan (Zurayk 2010). These, and other cities in the region, have managed to maintain areas within the city limits and the urban fringes exclusively for farming vegetable and cognate food despite pressures from rapid urbanization. Rami Zurayk (2010) has drawn attention to the vibrant urban agriculture industry of Amman, Jordan. According to Zurayk (2010), Amman is a busy city whose built environment is richly punctuated by many farming activities, including poultry, gardens, and grazing fields. These activities constitute not only a source of food for the urban centres; they are also the sole source of income for some in the region's poorer countries. The urban fringes serve mainly as a grazing site for goats and sheep although as Zurayk (2010) has observed, it is not unusual to find goats and sheep grazing in small green spaces within at the heart of cities in the region.

Private grazers occupy large areas around the semiarid steppe lands with population densities that are generally low but higher around irrigated areas. Such areas richly punctuate the lands, which by their arable nature help to support both a human and cattle population around major cities throughout the region. The cattle are mainly goats and sheep, and occasionally, camels. These are typically sold to large feedlots in the cities. The sale of these to large feedlots in the urban areas constitutes one link between rural areas and cities in the region. The Food and Agricultural Organization (FAO) (2015) has identified an additional link between the two locales. The link is located in the capital provided, and the control exercised over, pastoral herds, by urban centres. In addition to the herds, the nomadic groups that are wont to congregate around oases and irrigation schemes are involved in the production of dates, other palm specie, fodder and vegetable, the surplus of which they sell to urban residents. These are commonplace in Algeria, Tunisia, Libya, and Morocco (FAO 2015).

The last class of urban farms comprises those owned and operated by the state and/or municipal authorities. Governments in the MENA region are cognizant of the need for urban farming as an important element in communities desirous of promoting sustainable development. Apart from constituting a vital source of food, urban farming helps to achieve many social, economic and environmental goals. One of the social benefits of urban farming is that it provides many opportunities for people to know and interact with their neighbors. Its economic benefits include, among others, the fact that it provides a source of income for some urban residents.

The many environmental merits of urban farming can be appreciated in terms of its ability to maximize the utility of urban land, water, and other resources. For example, urban farming is helping countries such as Kuwait and the United Arab Emirates (UAE), which experience severe water scarcity problems save water in their food production processes.

However, the technology that permits the growing of crops with very little water is prohibitively costly. Accordingly, only the very wealthy countries in the MENA region have been able to operate large-scale urban farms. The case of the 'vertical farm' in Dubai in the United Arab Emirates described recently in a Cable News Network (CNN) online article is illustrative (CNN.com, Online). The vertical farm, which has an area of about three hectares, runs along the spine of the City of Dubai. It is a community farm that accommodates 38.000 square feet (about 3500 m^2) of 'Biodome Greenhouses' dedicated to the growing of fruits, herbs and vegetables. The farm contains more than 20 kinds of herbs and 2500 date palms, which are fertilized by green compost.

4.8 Energy in Built Space

As stated earlier, the Middle-East and North Africa (MENA) region is well endowed with energy resources, especially petroleum. However, a fact that is far less publicized about the region is its abundant reserves of renewable energy resources. One reason for this is that authorities in the region have just recently taken the steps necessary to maximize the utility of these resources. A few countries in the region have already embarked on efforts to incorporate renewable energy into their national energy plans. As Box 4.4 shows, Egypt is at the top of the list of countries in this category. A recent Working Paper by the University of Pennsylvania's Social Impact Research Experience (SIRE) research unit explores various renewable energy policies and legislative frameworks in the MENA region (see Abdelrahim 2019). Countries in the region are presently involved with initiatives to develop different renewable energy sources. The avowed aim of these initiatives is to develop sources to complement their rapidly dwindling conventional energy reserves. The other impetus for embarking on these initiatives are as follows: population growth; growing levels of industrialization; the need to reduce CO_2 emissions and mitigate the effect of climate change; and the need to counter the impact of dwindling reserves of hydrocarbons or fossil fuels in the region and globally.

The many sources of renewable energy that promise to be of enormous benefit to built space in the MENA region include solar, wind, and hydropower. Solar offers much promise for the region because it counts among the regions with the highest amount of sunlight on earth. Wind also ranks very high among the renewable energy sources with the most promise for the region. Because of its vast expanse of desert land, wind speeds in the region typically go as high as 13.8 m/s. However, this source remains relatively untapped; the few initiatives to develop the source have been limited to large-scale applications with hardly any directed at standalone wind

Box 4.4 Renewable Energy Initiatives in MENA: The Case of Egypt
Apart from being the home of the Aswan Dam, Egypt is also well-known in the region as a leader in efforts to incorporate renewable energy in its national energy generation and supply efforts. Well-endowed with oil, especially natural gas, Egypt has taken steps to maximize the utility of its abundant sunlight. The country has also moved to take advantage of the high wind speeds it experiences daily. The wind speeds often reach 13.8 m/s. Efforts in this regard are shepherded by the country's Ministry of Electricity and Renewable Energy (MOEE), which previously went under the appellation Ministry of Electricity. The name change was meant to underscore the country's commitment to renewable energy (RE). Another indicator of the country's commitment to RE is its ratification of the Paris Climate Change Agreement on June 29, 2017. More importantly, the country, through its Ministry of Electricity and Renewable Energy (MOEE), released the Integrated Sustainable Energy Strategy to 2035—in other words, ISES 2035. This details the country's commitment to throw its full weight behind renewable energy and sustainable development writ large. As per ISES 2035, the Egyptian Ministry of Electricity and Renewable Energy is committed to supplying 20% of the country's energy from renewable sources by 2022.

Source: Abdelrahim (2019).

turbines (Abdelrahim 2019). Although typically treated as a conventional energy source, hydropower also constitutes an important renewable energy resource in the MENA region. An environmentally friendly source because it does not interfere with the natural flow of the earth's water, hydropower has been a source of energy in the region for a long time. However, it has depended mainly on large dams such as the Egyptian Dam or Aswan Dam.

4.9 Conclusion

The MENA region is unique in many ways. In analyzing the region in this chapter, I have attempted to promote understanding of the factors that affect efforts to promote nature in built space. This latter tends to differ by geo-cultural circumstances. In most cases, this would be taken to comprise structures of concrete or other solid materials and paved streets. For inhabitants of nomadic communities, it comprises not only these familiar structures but also the makeshift shelters, ancillary facilities and the foot tracks connecting their villages to the communal fields that serve as the food source for their livestock. This chapter has analyzed key socio-cultural, politico-economic, and techno-ecological drivers of policies and other actions focusing on these environmental matters. The unique nature and importance of these features are discussed with respect to five specific substantive environmental matters of utmost importance to the built environment. These include urban land, urban forestry, green architecture, water resource management, energy and food. The chapter has demonstrated that efforts to promote these elements of nature in built space have been affected by political, economic, social, ecological, cultural, and historical factors that constitutes defining features of MENA.

References

Abdelrahim F (2019) The rise of renewable energy in the MENA region: an investigation into the policies governing energy resources. Working paper, Social Impact Research Experience (SIRE), The University of Pennsylvania

Akerlund U, Knuth L, Randrup TB, Schipperijn J (2006) Urban and peri-urban forestry and greening in West and Central Asia: experiences, constraints and prospects. LSP working paper 36: access to natural resources sub-programme, Food and Agricultural Organization (FAO)

Barnouw V (1961) Chippewa social atomism. Am Anthropol 63(5):1006–1013

Barrett R, Leal J (2017) Planning for successful street trees in arid climates. Online article on deeproot.com. Accessed 12 July 2019 at: http://www.deeproot.com/blog/blog-entries/planning-for-successful-street-trees-in-arid-climates

Barton A (2019) Water in crisis – Middle East. Online article for The Water Project, Inc. Accessed 2 July 2019 at https://thewaterproject.org/water-crisis/water-in-crisis-middle-east

Beeson M (2010) The coming of environmental authoritarianism. Environ Polit 19(2):276–294

Bento G, Dias G (2017) The importance of outdoor play for young children's healthy development. Porto Biomed J 2(5):157–160

Brockopp J (2012) Introduction to special issue: Islam and ecology, law and practices of Muslim environmentalism. Worldviews 16:213–217. https://doc-research.org/our-events/economic-social-development-middle-east-north-africa-success-failure/

CNN.com (Online) World's largest vertical farm to feed Middle East's high-fliers. Accessed 30 July 2019 at: https://www.cnn.com/travel/article/dubai-vertical-farm-emirates-catering/index.html

Dixon J, Gulliver A, Gibbon, Hall M (2001) Farming systems and poverty: improving farmers' livelihoods in a changing world. Rome: FAO. Available Online. Accessed 27 July 2017 at: www.fao.org/3/Y1860E/y1860e05.htm

DOC (2019) Economic and social development of the Middle East and North Africa: success or failure? Article by the Dialogue of Civilization Research Institute. Available online. Accessed 10 July 2019 at: https://doc-research.org/aug19/our-events/economic-social-development-middle-east-north-africa-success-failure/

Doyle T, Simpson A (2006) Traversing more than speed bumps: green politics under authoritarian regimes in Burma and Iran. Environ Polit 15(5):750–767. https://doi.org/10.1080/09644010600937199

Draz O (2019) The hema system in the Arabian Peninsula. Retrieved, July 7, 2019 from: http://www.nzdl.org

England A (2019) The 13 best heat-tolerant trees. Online Article on: The Spruce. Accessed 12 July 2019 at: https://www.thespruce.com/best-heat-tolerant-trees-2539920

Fahmy A, Dhillon N (2008) Urban and young: the future of the Middle East (Op-Ed). Available online. Accessed 10 July 2019 at: https://www.brookings.edu/opinions/urban-and-young-the-future-of-the-middle-east/

FAO (2002) Near east forestry commission: the use of Treated Waste Water (TWW) in forest plantations in the near east region. Secretariat Note. Available online. Access on 26 July 2019 at: http://www.fao.org/DOCREP/MEETING/004/AB400E.HTM

FAO (2015) Status of the world's soil resources: main status. (Chapter 13: regional assessment of soil changes in the Near East and North Africa). Rome: Italy. Available online. Retrieved, July 29, 2019 from: http://www.fao.org/3/a-bc602e.pdf

FAO (2019) An overview of land tenure in the near east region. (Part 1). Report produced for FAO (Food and Agriculture Organization) by Jon Rae. Available online. Accessed 10 July 2019 at: http://www.fao.org/3/a-aq202e.pdf

Faruqui NI (2001) Islam and water management: overview and principles. In: Faruqui NI, et al (eds) Water management in Islam. United Nations University Press, Tokyo

Faraqui NI, Biswas AK, Bino MJ (2001) Preface. In: Faraqui NI, Biswas AK, Bino MJ (eds) Water management in Islam. United Nations University Press, Tokyo

Faraqui N, Niang S, Redwood M (2002) Untreated wastewater reuse in market gardens: a case-study of Dakar, Senegal. In: Paper presented at the international water management institute workshop on wastewater reuse in irrigated agriculture: confronting the livelihood and environmental realities, international water management institute. Hyderabad, India

Gasiorowski M, Yom SL (eds) (2017) The government and politics of the Middle East and North Africa. Westview Press, Boulder

Gertel J, Hexel R (eds) (2018) Coping with uncertainty: youth in the Middle East and North Africa. Saqi Books, London

Ghazouani W, Molle F, Rap E (2012) Water users associations in the NEN region: IFAD interventions and overall dynamics. Draft report submitted to IFAD and the International Water Management Institute (IWMI). October 2012

Gilley B (2012) Authoritarian environmentalism and China's response to climate change. Environ Polit 21(2):287–307

Ginsburg KR (2007) The importance of play in promoting healthy child development and maintaining strong parent–child bonds. Pediatrics 119:182–188

Hashem N (2013) Assessing spatial equality of urban green spaces provision: a case study of greater Doha in Qatar. Local Environ 20(3):386–399. https://doi.org/10.1080/13549839.2013.855182

Heilbroner RL (1974) An inquiry into the human prospect. Norton, New York

Hope M, Young J (2019) Islam and Ecology. Online publication. Accessed 5 July 2019 at: http://www.crosscurrents.org/islamecology.htm

Imam.org (2017) How involve do I need to be in my community? Online publication, July 10, 2017. Accessed 7 July 2019 from https://www.imam-us.org/involved-need-2/

Kjerfve B (2014) The state of higher education in the Middle East: documenting change, challenges and an undeniable opportunity. An Op-ed in U.S. News & World Report Online, Dec. 17, 2014. Accessed 12 July 2019 at: https://www.usnews.com/news/college-of-tomorrow/articles/2014/12/17/op-ed-the-state-of-higher-education-in-the-middle-east

McBride J, Mossadegh A (2000) Tree-lined canals and the urban forest of Tehran. Arboric J 24:155–173

Nasr SH (1968a) Man and nature: the spiritual crisis of modern man. George Allen and Unwin Ltd., London

Nasr SH (1968b) Science and civilization in Islam. Harvard University Press, Cambridge

Nasr SH (1972) Sufi essays. George Allen and Unwin, London

Nasr SH (1976) Islamic science: an illustrated study. World of Islam Festival Publishing Co/Thorson Publishers, London

Nasr SH (1989) Knowledge and the sacred, *t*he 1981 Gifford lectures. State University of New York Press, Albany

Okuma N, Shinoda T (2009) Expansion of water recycling business in Dubai, Middle East. Hitachi Rev 58(6):246–250

Rae J (2002) An overview of land tenure in the near east region. Part I; Part II, individual country profiles; and Bibliography. Rome, FAO. (mimeo)

Scherer D (1982) Anthropocentrism, atomism, and environmental ethics. Environ Ethics 4(2):115–123

Silva CS, Viegas I, Panagopoulos T, Bell S (2018) Environmental justice in accessibility to green infrastructure in two European cities. Land 7:134. https://doi.org/10.3390/land7040134

Sowers J (2007) Nature reserves and authoritarian rule in Egypt: embedded autonomy revisited. J Environ Dev 16(4):375–397

Tawk ST, Moussa Z, Hamadeh SK (n.d.) Mainstreaming urban agriculture in the Middle East and North Africa: a multi-stakeholder approach. Available online. Accessed 23 July 2019 at: ifsa.boku.ac.at/cms/file.admin/Proceeding2014/ws_2_8_Tawk.pdf

The Economist (2015) Islam and ecology: in almost perfect harmony. Retrieved, July 5, 2019 from: https://www.economist.com/erasmus/2015/08/21/in-almost-perfect-harmony

Topcu M, Girgin S (2016) The impact of urbanization on energy consumption in the Middle East. J Int Global Econ Stud 9(1):21–28

UN Environment (2018) How Islam can represent a model for environmental stewardship. Accessed 7 July 2019 at: https://www.unenvironment.org/news-and-stories/story/how-islam-can-represent-model-environmental-stewardship

UNDESA (2019) World urbanization prospects 2018 revision. United Nation Division of Economic and Social Affairs (UNDESA). Available online. Accessed 10 July 2019 at: https://www.un.org/development/desa/publications/2018-revision-of-world-urbanization-prospects.html

UNEP (2012) Global environmental outlook (GEO-5): environment for the future we want. United Nations Environmental Programme, Nairobi

UNICEF (2019a) MENA generation 2030: investing in children and youth today to secure a prosperous region tomorrow. Online publication of the United Nations Children's Fund (UNICEF). Accessed 10 July 2010 at: https://www.unicef.org/mena/reports/mena-generation-2030

UNICEF (2019b) Social policy: challenges. Online Publication by the United Nations Children's Fund (UNICEF). Accessed 12 July 2019 at: https://www.unicef.org/mena/social-policy

USAID (2016) Land tenure in urban environments. Issue brief. Available online. Accessed 16 July 2019 at: https://www.land-links.org/wp-content/uploads/2016/09/USAID_Land_Tenure_Urban_Brief_061214-1.pdf

Uyl DJ, Rasmussen DB (2006) The myth of atomism. Rev Metaphys 59(4):841–868

WCPA (2019) WCPA North Africa, West Asia and the Middle East. Online article for the World Commission on Protected Areas (WCPA). Retrieved, July 4, 2019 from: https://www.iucn.org/commissions/world-commission-protected-areas/regions/wcpa-north-africa-west-asia-and-middle-east

Wells P (2007) The green junta or, is democracy sustainable? Int J Environ Sustain Dev 6(2):208–220

WHO (2019) Ending childhood obesity. Geneva: World Health Organization; 2016. Available from: http://www.paho.org/bra/index.php?option=com content&view=article&id=4997:relatorio-da-comissaopelo-fim-da-obesidade-infantil-busca-reverter-aumento-de-sobrepeso-eobesidade&Itemid=821

Wily LA (2018) Collective land ownership in the 21st century: overview of global trends. Land 7(2):68. https://doi.org/10.3390/land7020068

World Bank (2019) Middle East and North Africa. Available online. Accessed 10 July 2019 at: http://web.worldbank.org/archive/website01418/WEB/0__CO-49.HTMhttp://web.worldbank.org/archive/website01418/WEB/0__CO-49.HTM

Youthpolicy.org (2019) Middle East and North Africa: youth facts. Available online. Accessed 10 July 2010 from: http://www.youthpolicy.org/mappings/regionalyouthscenes/mena/facts/

Zurayk R (2010) From incidental to essential: urban agriculture in the Middle East. J Agric Food Syst Commun Dev 1(2):13–16

Chapter 5
Nature in Built Space in Asia and the Pacific

Abstract The Asia and Pacific region has been experiencing many problems arising from heightened levels of urbanization, population growth, and industrialization. These factors have exacerbated many problems in built space, especially those relating to energy, food and water shortages as well as traffic congestion and air pollution. Paradoxically, these problems have been compounded by the unprecedented levels of economic growth experienced in the region in the recent past. For instance, problems of traffic congestion, noise and air pollution were relatively absent from the region until the dawn of its 'economic miracle.' This chapter has shed some light on efforts to promote nature in built space in the face of the identified problems. In particular, the chapter has discussed the many ways by which the region's political, economic, social, technological, ecological, cultural and historical contexts have affected initiatives to create and maintain forests, gardens, parks, and green infrastructure in general in its built space.

5.1 Introduction

The focus of this chapter is the Asia and Pacific region. For international politico-administrative purposes, the region contains the following sub-regions: Central Asia, North-East Asia, South Asia, South-East Asia and the Pacific. Table 5.1 shows the specific countries in each of these sub-regions. Geographically, it extends from the Pacific Basin to the South China Sea, and from the Indian Ocean to the Antarctica.

The executive summary of the publication, *A Long-term Perspective on Environment and Development in Asia-Pacific* by ECOASIA contains very useful introductory material on this region (see ECOASIA Online). The region covers a vast area encompassing a diverse ecosystem dominated by marine life. It lays claim to a very rich history, an array of cultural and ethnic groups. As stated in previous chapters, this book assigns much importance to culture as a determinant of people's relationship with the environment. By extension, we consider religion, and culture in general, an important determinant of attitudes towards nature in built space. The cultural context of the Asia and Pacific region, like its other attributes, is relatively diverse. However, there are a number of common threads that tie groups of countries within the region together. For example, the Confucian ethics is as much an attribute of Chinese and Korean culture as it is Japanese. Similarly, the Ramayana

A. J. Njoh, *Nature in the Built Environment*,
https://doi.org/10.1007/978-3-030-39759-3_5

Table 5.1 Countries in Asia and Pacific by sub-region

Sub-region	Countries
1. Central Asia	Armenia, Azerbaijan, Georgia, Kazakhstan, Kyrgyzstan, Tajikistan, Turkmenistan, Uzbekistan.
2. North-East Asia	China, Japan, Democratic People's Republic of Korea, Russian Federation, Mongolia.
3. South Asia	Afghanistan, Bangladesh, Bhutan, India, Islamic Republic of Iran, Maldives, Nepal, Pakistan, Sri Lanka, Turkey.
4. South-East Asia	Brunei Darussalam, Cambodia, Indonesia, Lao People's Democratic Republic, Malaysia, Myanmar, Philippines, Singapore, Thailand, Timor-Leste, Viet Nam.
5. Pacific	Australia, Cook Islands, Fiji, French Polynesia, Guam, Kinbati, Marshall Islands, Federated States of Micronesia, Nauru, New Caledonia, New Zealand, Niue, Northern Mariana Islands, Palau, Papau New Guinea, Samoa, Solomon Islands, Tonga, Tuvalu, Vanuatu.

is as much a part of Indian culture as it is Indonesian. As for religious affiliation, Armenians who are geographically separated from Australians and New Zealanders are connected by their avowed Christian faith. In an identical manner, the politically, economically and demographically disparate polities of Afghanistan, Pakistan, Bangladesh and Indonesia are linked by Islam. Finally, Bhutan, and Cambodia find common ground in Buddhism.

Environmentally, the region has been experiencing several problems, many of which arise from heightened levels of urbanization, population growth, and industrialization. These factors have conspired to exacerbate its energy problems. For one thing, the rate of energy consumption is faster than the rate of energy generation. Industrialization appears to be a mixed blessing for the region. While it has engendered much needed economic growth, it has also been incriminated as the leading source of air, noise, and water pollution. It is at the root of other increasingly nagging environmental problems such as toxic chemical pollution in the region. In addition, cities in the region, because of improved socio-economic conditions, have been experiencing unprecedented levels of traffic congestion and commensurate environmental problems such as urban air pollution. Cities such as Shanghai in China, Tokyo in Japan, Jakarta in Indonesia and New Delhi in India have been wrestling with traffic congestion and its attendant problems for decades. The increasing severity of the problem accentuates the need to be more attentive to the nature in the built environment question throughout the region.

This chapter is dedicated to promoting understanding of how the region's politico-economic, socio-cultural, geo-ecological and historical contexts facilitate or inhibit initiatives to create and preserve nature in built space. It continues in the next section by painting a vivid picture of the region's political, economic, social, technological, ecological, cultural and historical (PESTECH) contexts. Then, it discusses religion, an important cultural element as a significant determinant of environmental stewardship. Then, it proceeds in subsequent sections to discuss efforts to develop and preserve specific elements of nature such as trees and related vegetation, food, energy and water in built space.

5.2 Geo-ecological, Politico-Economic and Historic Contexts

The Environmental Congress for Asia and the Pacific (ECOASIA) has painted a vivid picture of the geo-ecological attributes of the Asia-Pacific region (see ECOASIA Online). The Asia and Pacific region is considerably large; it is the most biodiverse region in the world. It covers a vast geographical area that stretches from Mongolia in the north to New Zealand in the south; it spans from the island states of Oceania in the east to Pakistan in west. The climatic conditions vary widely, and range on a continuum from tropical on one end, to temperate on the other. The topographical features of the region are also very diverse; they range from the highest peak of the Himalayas to the lowest point at the deepest ocean floor in the Sulu Sea. The region contains three main bio-geographical sub-regions, including the continental, the archipelagic and small island ecosystems.

The Asia-Pacific region contains politico-economic systems that include liberal democracies such as Australia, and New Zealand, semi-free market economies under highly centralized politico-administrative systems such as China, and centrally-controlled politico-economic systems such as North Korea. Thus, central governments tend to play an active role in the region's economy. By the same token, central governments are exceedingly active in policy formulation and implementation. This is especially true in the case of environmental policy. Promoting nature in built space falls within the orbit of environmental policy. Thus, it follows that urban natural environmental policy making and implementation are the responsibility of central governments in the region. The problem, however, is that ensuring the development and preservation of nature in built space constitutes a local issue. Ideally, therefore, it must be addressed by local, and not national level authorities.

China exemplifies countries with powerful central governments; it plays an active role, and with great success, in the economy. In fact, China's success record challenges the notion that political democratization and economic liberalization constitute the one best way to attain socio-economic development. The region has registered the highest growth rate during the last three decades. Yet, about half of the region's population lives on less than one dollar a day. Another noticeable trend in the region has been on the demographic front. Here, it is necessary to reiterate the fact that the region is home to 60% of the world's population. In addition, it is home to two of the world's most populous nations, China and India. Paradoxically, the region is also home to some of the world's smallest sovereign nations; these include the smallest Pacific Island countries. Questions concerning nature in built space in such countries invariably assume national as opposed to local importance. Other noticeable demographic changes in the region include declining rates of fertility and mortality; however, this tends to vary significantly across the region. For example, while the regional average fertility rate is 2.1 births per woman, the rate stands at

1.7 in East Asia, and 2.5 children per woman in the south (UNFPA Online). This has resulted in a net population gain, hence, rapid population growth in the region as a whole. The trend has also resulted in a 'youth bulge' in the South Asia sub-region. Thus, this sub-region boasts a youthful population that can contribute to developing, and constitute an effective demand for, green and related infrastructure in built space. However, the tenability of this proposition depends on the state of the region's economy. The East Asia sub-region, which has been registering low birth rates contain an ageing population. Thus, such areas are expected to experience difficulties provisioning, utilizing and maintaining green and related infrastructure in built space.

It is important to underscore the fact that unlike regions in the developing world, this has been accompanied, if not triggered by, significant economic growth. This has been ongoing for more than two decades. This economic growth, which has been the highest in the world, is expected to continue into the foreseeable future. One outcome of this growth has been heightened intra-regional trade. Countries in the region are at different levels of economic development. Thus, for example, while China, Australia, New Zealand, Japan and South Korea are economically well developed, others such as Cambodia, Bangladesh, Vietnam, and Papua New Guinea qualify as typical developing economies.

Implications of Recent Trends As stated earlier, the Asia and Pacific region has been experiencing some notable demographic, social and economic changes since the last two or so decades. The changes have far-reaching implications for efforts to create and maintain nature in built space in the region. Some of the changes pose a direct threat to the natural environment, and by extension, nature in built space. For example, the rapid and unprecedented rates of population growth and urbanization have led to unsustainable agricultural practices in the region. Rapid population growth and urban physical expansion have also led to the overuse of natural resources, and deforestation. Deforestation has in turn led to soil infertility, erosion and water salination, which constitutes a grave threat to agricultural activities in the region. These problems accentuate, rather than diminish, the need for green infrastructure such as forests, greenery, gardens and parks in built space throughout the region.

Water salination, a problem that has already been observed in Australia, constitutes an enormous threat to freshwater in the region. Generally, the region faces water problems of both the qualitative and quantitative variant. To be sure, the problem does not approximate what obtains in the Middle East and North Africa. Yet, the region's problem of limited access to freshwater has risen to the level that has caught the attention of analysts and international organizations such as the United Nations Environmental Programme (UNEP). The importance of in the built environment, particularly urban areas, cannot be overemphasized. Freshwater is necessary not only for domestic use, but also for irrigation and industrial purposes. This problem is especially acute on the region's many islands, some of which, despite their diminutive sizes, are sovereign nations. Rapid population growth, increasing urbanization and industrialization as well as greenhouse emitting activities, are also exacerbating climate change and related problems. These problems threaten not only nature in built space, but entire built areas in fragile locales such as low-lying islands in the region. These islands risk being overwhelmed and completely effaced out of existence by rising sea levels. Climate change is also increasingly threatening marine habitats such as coral reefs and mangroves in the region. Also, there has been an increase in the frequency of severe weather events in the region. Many more of these events are highly likely in the future thanks to observable changes in the region's climate. Significant increases industrial activities and use of motorized transportation have already been incriminated as dominant sources of carbon dioxide (CO_2 emissions). This exemplifies a situation in which urban based activities have negative implications both for nature in built space in particular, and the natural environment as a whole.

The region's gains in the industrial and economic milieu appear to be a mixed blessing. These gains, which have occurred mainly in urban areas, have boded ill for efforts to promote the region's biodiversity. In particular, the gains have led to resource overexploitation, habitat fragmentation and above all, environmental degradation. Industrial activities, their myriad advantages notwithstanding, have had many negative externalities for both the built and natural environment in the region. One of these externalities is chemical waste. One of the nagging problems in this regard relates to the disposal of this waste, as well as reclaiming the sites on which such waste has ever been disposed. Another serious negative externality of industrialization in the region is air pollution. Here, it is worth noting that, China, the region's largest economy relies mainly on coal, which is arguably the largest emitter of carbon dioxide (CO_2). Rather than decrease its dependence on coal, the country actually increased its coal production by 3–5% from 2005 to 2009 (UNEP 2016). Natural gas, by far, a cleaner energy source, constitutes only 10–25% of the country's energy mix; only 4% of its energy need is met by natural gas.

The implications of the region's rapid industrial, economic and demographic growth for nature in the built environment are both positive and negative. Consider the region's impressive economic growth record of the last couple of decades. The value of this trend can be best understood by appreciating the cost of developing and maintaining green systems such as urban trees, parks and gardens. The cost of these systems is neither inconsequential nor trivial. An accurate computation of the cost of such systems must account for the time and resources necessary for their maintenance. Such a computation must also be cognizant of the opportunity cost of the land and other space taken up by the systems. As a specific example, let us consider the cost of urban trees. A recent extended review of the relevant literature has highlighted the need to appreciate the cost of urban trees; these are often promoted as a viable means for improving urban quality of life, lowering carbon dioxide emission, and cleaning urban air (Song et al. 2018). The review breaks the cost of urban tree planting and maintenance into two broad categories, namely direct and indirect costs. The direct cost includes mainly the cost of planting and maintaining the trees. The indirect costs include expenditures associated with damages to buildings caused by tree roots, injury and/or damage from falling trees, disruptions to traffic occasioned by periodic maintenance work on the trees. Research on urban forests and urban greening in general have typically focused on the developed world.

Consequently, there is a dearth of knowledge on the cost of urban green infrastructure in developing countries.

Accordingly, to gain some understanding of the cost of urban trees, we must depend on studies that have developed countries as empirical referent. One study by Escobedo and Seitz (2009) comes to mind. According to this study, the national mean annual expenditure on a public tree in the United States in 2007 was $10.62 (Escobedo and Seitz 2009). The study also focuses on the cost of maintaining urban forests to municipalities. This cost is estimated to be approximately 0.5% of the annual budget of an average municipal government in the U.S. (Escobedo and Seitz 2009). An itemized list of this cost would include the value of items such as tree acquisition, planting, irrigation, fertilizers, leaf blowing and pruning. Based on the U.S. experience, urban forests maintenance expenditure as a proportion of the annual budget is higher for larger than smaller municipal governments (Escobedo and Seitz 2009). This is likely a function of the fact that smaller municipalities are more likely than their larger counterparts to benefit from voluntary labor input from residents. This input is free of charge to the municipalities although it is not completely cost-free. The cost of such labor input is quite significant. For example, homeowners in Orlando, Florida spend on average 8 h per year blowing leaves, hedge-trimming and tree pruning (Escobedo and Seitz 2009). In 2007, the City of Gainesville, Florida, spent $1,559,952 on its urban forests (Escobedo and Seitz 2009., p. 2). This translated to about $10.57 per public tree.

The region is well represented by countries at both ends of the economic and industrial development continuum. This means that both labor-intensive and capital-intensive strategies can be effectively adopted in efforts to promote nature in built space in the region. Countries such as China, Japan, South Korea, Australia and New Zealand are capable of adopting plans that have proved useful in urban greening initiatives in industrialized and economically advanced countries such as those of Western Europe and North America. The less economically successful economies such as Bangladesh, Cambodia, and Afghanistan would do well to employ labor-intensive strategies that call for the labor input of citizens. Such a strategy would be in line with principles of community participation.

5.2.1 Culture, Beliefs and the Natural Environment in Asia and the Pacific

The Asia-Pacific is a vast region that includes populous and densely populated countries such as India, China, South Korea, Thailand and Japan, city-states such as Hong-Kong and sparsely populated countries such as Australia. To the extent that religion constitutes the most prominent aspect of culture, it is safe to state that the Asia and Pacific region is, to a certain extent, culturally diverse. Thus, it is logical to assert that people's relationship with the natural environment in the region is influenced by the doctrine of the dominant religion in each country. The countries have different dominant religions, and in some cases, they have state religions. For instance, Islam is the state religion of Pakistan while Christianity is the state religion of Armenia. Armenia is unique in this case because it is the first country—since 301 AD—to make a branch of Christianity, the Armenian Apostolic Church, the state religion.

The Pew Research Centre records Hindus as constituting the largest religious group in the A-PAC region (PRC 2015). Hinduism boasted as many as one billion adherents in 2010 (PRC 2015, para. 2). Muslims constitute a close second, and are projected to surpass the population of Hindu adherents by 2050. Another group with a significant and growing number of adherents is Buddhism. Almost all (97%) of Cambodians are Buddhist. Christians constitute a small minority (7%) of the population in the Asia and Pacific region. The chapter on the Middle East and North Africa in this book discusses in some detail the relationship between Islam and the natural environment. The chapters on North America and Western Europe respectively, discuss the relationship between Christianity and environmental stewardship. Therefore, no more than a cursory examination of how Christianity and Islam affect their adherents' views of the environment is necessary here.

5.2.2 *Christianity, Islam and Environmental Stewardship*

Critics typically criticize the manner in which Christianity and Islam view the natural environment as inherently anthropocentric. From this vantage point, Islam and Christianity are seen as placing human beings as not only separate from, but also superior to, nature. Above all, within the framework of Christian and Islamic doctrine, critics contend that human life possesses intrinsic value; in contrast, other creatures exist to be exploited by, and for the upkeep of, humans. There is no shortage of passages in *Qur'an* and *Hadith* heralding the Almighty God's wonderful creations, and his charge to humans to be in charge of these. The role of humans in this connection is that of trustee *(khalifa)* as found in Qur'an 2:30. Similarly, many passages in the Bible beseech Christians to be good custodians of the natural environment. For example, Genesis 2:15 states thus:

> "The Lord God took the man and put him in the garden of Eden to work it and keep it." On its part, Numbers 35: 33–34 states that "You shall not pollute the land in which you live, no atonement can be made for the blood that is shed in it. You shall not defile thee land in which you live, in the midst of I dwell."

Yet, critics would argue, Christianity and Islam are not eco-friendly, and can therefore not be depended upon for inspiration to promote nature in built space. Many Biblical passages do not help the case for Christianity as an environmentally-friendly religion. For instance, Genesis 9: 3 states thus: "Every moving thing that lives shall be food for you. And I as give you the green plant, I give you everything." The Qur'an contains similar passages. Yet, many an Islamic philosopher would fervently insist that Islam calls for environmental stewardship and nothing less.

Prominent among those articulating this point of view is *Seyyed Hossein Nasr (see e.g., Nasr 1968a, b, 1972, 1976, 1987, 1989). He contends that Islam assigns to humans the responsibility of caring for the natural environment. In summarizing this viewpoint, Hope and Young (2019, para. 1) state that Nasr "sees at the center of Islam a charge to protect the natural world—a world that reflects the higher reality of the transcendent God."* Thus, it is safe to conclude that Islamic doctrine not only recognizes the importance of nature; it actually implores its adherents, Muslims, to actively engage in protecting it. In this regard, Muslims are beseeched to respond to environmental threats as part of Islam. The Qur'an instructs humans to (Hope and Young: para. 4): "'Do no mischief on the earth after it hath been set in order, but call on him with fear and longing in your hearts: for the Mercy of God is always near to those who do good' (Q.7:56)." Therefore, contrary to the views of critics, Islamic doctrine is of great utility to efforts to create and manage nature in built space.

5.2.3 Hinduism and Environmental Stewardship

Hinduism, the religion with the largest number of adherents in the Asia and Pacific region, echoes the sentiment of Islam with respect to environmental stewardship. An ancient Hindu adage is exceedingly telling in this regard. It states thus: "The Earth is our mother and we are all her children" (BBC Online). Also underscoring the importance of environmental stewardship in Hindu philosophy, Mathama Gandhi once said, "there is enough for everyone's need but not enough for their greed" (BBC Online). The need to treat the environment with care—in other words, environmental stewardship, is at the heart of Hinduism. The word environment according to Hindu doctrine connotes the natural world. This means "everything around us that is part of the Earth and nature" (BBC Online, para., 1). Hindu doctrine contains key teachings about the Earth and nature writ large. Foremost along these lines are Ahimsa, Karma, and Moksha.

Ahimsa, the principle of non-violence, teaches Hindu adherents that every living thing is sacred because it was created by God. Non-violence, as a principle in this case beseeches Hindus to respect all life forms. A second principle is Karma, which requires adherents of Hinduism to recognize that all actions have consequences. Therefore, it is important for people to refrain from actions intended to harm any living thing, that is, any of God's creation. This is because the consequences of such actions are to be felt by the perpetrator either in the present lifetime or any other lifetime. This latter is possible in Hindu religious thought thanks to the belief in reincarnation. A related concept in Hindu philosophy is that nature and humans are inextricably interconnected. Therefore, any harm to nature invariably harms humans. Thus, "nature cannot be destroyed without humans also being destroyed" (BBC Online, para. 4). Finally, there is the concept of Moksha, which is the belief in reincarnation. Thus, death does not constitute the end of human life. Rather, it is the end of one life that permits the start of another—reincarnation. However, Hinduism holds that at some point in life, a human is released from the cycle of reincarnation. It is at this point, it is believed, that all suffering for that human being ends. Thus, Hindus view the act of righteousness towards, all living things as the starting point in the quest for Moksha.

Apart from Islam and Hinduism, which are well-known outside the Asia and Pacific region because of the large size of their adherents, the region boasts many other, albeit, smaller religions. These are mainly religions that are indigenous to the region and have not spread to other parts of world. A few of these, including Shinto (mainly in Japan) and Buddhism (mainly in Cambodia) have well developed philosophies on the natural environment. The environmental philosophy of aboriginals in Australia and New Zealand provide an informative backdrop for a meaningful discussion of environmental stewardship as an element of indigenous religious thought in the Asia-Pacific region, it helps to consider the case of aboriginals in Australia.

5.2.4 Aboriginals and Environmental Stewardship

The aboriginals of Australia and New Zealand have a rich and sophisticated indigenous culture and belief system complete with a philosophy of environmental stewardship. With its emphasis on reverence for land, the system is similar to other indigenous religions. In this regard, it depends on oral traditions to ensure continuity. However, it is important to note that aboriginals do not constitute a monolithic group. Rather, they are comprised of multiple sub-groups, each with a few unique features of its own. Yet, they share some common characteristics, especially with respect to beliefs and doctrines. For example, all aboriginal groups believe in life after death.

The practices typically incorporate aspects of the natural environment. One example is the fire-stick farming, which is a practice involving the systematic burning of portions of vegetation to facilitate hunting and above all, reduce the incidence of bush fires. Considered as good husbandry, this practice also helps to fertilize the land and increase the food inventory of edible wildlife such as kangaroos. Another common religious practice is the walkabout, which requires adolescent boys to undergo a rite of passage by exiting and living away from their family group area. Also, the various groups share the belief in ancestral beings. One such being that is revered by aboriginals without exception is the Rainbow Serpent. Another is Banjame, also known as Bunjil, which is believed to be the primary creator-spirit in South-East Australia. Then, there is Dingo Dreaming, an important ancestor. Finally, there is the priority place afforded traditional healers, who go under the name, *Ngangkari* in the country's central region. Traditional healers or doctors are also expected, as in African and other indigenous ethos, to serve as the custodians of the society's history, culture and important stories.

5.2.5 Shinto and Environmental Stewardship

One important indigenous religion that is seldom known outside of the region is Shinto. It is practiced mainly in Japan and does not operate neither worshiping houses nor lay own any congregations. Until the 1950s, Shinto served as the state religion of Japan. Like other indigenous religions, it also emphasizes reverence for nature. In this connection, it preaches that the power of Kami, an entity believed to exists between God and spirits can be invoked ritualistically to assist Shinto adherents in times of difficulties. Additionally, it operates many sacred places, including shrines, which are usually set in beautiful gardens. These gardens are believed to venerate the spirits of nature. Also, it emphasizes hygiene and sanitation. Adherents are required to thoroughly clean themselves before praying or invoking deities.

An understanding of indigenous religion is important not only for its own sake; rather, it is necessary to provide some contrasts with organized religions and their environmental philosophies. Although throughout history, organized religions have derided indigenous religions as primitive, polytheistic and ignorant writ large, there is abundant evidence demonstrating their environmental consciousness. One of the most persuasive arguments tying indigenous religion to environmental stewardship has been advanced by John Grim (2019). He begins his paper on "Indigenous traditions and ecology" by defining the term 'indigenous' as "a generalized reference to the thousands of small-scale societies who have distinct languages, kinship systems, mythologies, ancestral memories, and homelands" (Grim 2019, para. 1). Paradoxically, despite the fact that these societies comprise at least 200 million people, their religions do not count among 'world religions' such as Buddhism, Islam and Christianity. The reasons for this paradox are beyond the scope of this book. Suffice to state that they share one trait that undeniably heightens their appeal to environmentalists. They are intensely attentive to nature. Grim (2019) proffers four themes capable of facilitating appreciation of the link between indigenous religions and ecology, including, kinship, spatial and biographical relations with place, traditional environmental knowledge, and cosmology. As a theme in this discourse, kinship highlights a set of important traits—emphasis on the integrity of all reality and the maintenance of intimate relations with the natural world—that are commonplace in indigenous societies. Evidence of these relations can be found in the lingo and known tales in these societies. It is not unusual to find the personification of animate as well as inanimate objects in indigenous languages.

The Lakota and their linguistic cousins, the Nakota and Dakota in the United States, for instance, speak of 'deer peoples' and 'bird nations.' Thus, the notion of kinship in such societies is not limited to human kin. Rather, it includes as is the case of Gwich'in peoples of Alaska, the concept of kinship encompasses all life forms (Grim 2019). Yet, what is remarkable here is not simply the fact that kinship extends to all life forms; it is the intellectual view that no being—not even the human being—is superior to other species. Here, as previously noted, indigenous mythology can be invoked to demonstrate the veracity of this seemingly absurd assertion. Drawing again from North America, Lakota culture personifies rocks and stones. In fact, there are many oral accounts in this culture that convey the notion that stones occupy an important place in the sequence of creation. Perhaps more cynical, when viewed through the analytical prism of a Westerner, is the fact that indigenous belief systems assume that everything constituting part of the natural world is endowed with knowledge and feelings. This belief invariably instills in indigenous people a level of environmental awareness that would be impossible otherwise. Thus, as John Grim (2019) noted, "Care for the earth is woven into the governance systems of indigenous people."

The Yekuna of Venezuela have regulated the amount of materials used for basket-weaving for centuries. If nothing else, this demonstrates sophisticated knowledge of the finite nature of natural resources and the need for conservation. These societies, the missionaries recounted had gods for almost everything under the sun—gods of rivers, gods of forests, gods of mountains, gods of prairies, and so on. These gods were revered and deterred humans from doing harm to the natural objects they protected. Thus, reverence for the 'river god' deterred humans from doing harm to rivers; while the 'forest god' deterred humans from harming the forest. Christian missionaries considered this sacrilege and moved speedily to supplant these beliefs and the commensurate practices with Christian doctrine.

5.2.6 Buddhism and Environmental Stewardship

Another major religion with roots in Asia, but less known beyond eastern and central Asia is Buddhism. It is a product of the teaching of Siddhartha Gautama. Concern for nature is said to be of centrality in Buddhist doctrine; in fact, attention to Buddhism and ecology has been experiencing gradual growth during the last two or so decades (Clippard 2011). One of the best-known scholarly works underscoring the strong bond between Buddhism and nature, particularly the natural environment is the edited volume by Alan Hunter Badiner under the caption, *Dharma Gaia: A Harvest of Essays in Buddhism and Ecology,* which was initially released in 1990 (see Badiner 1990). The book, especially the essay by Thich Nhat Hanh, goes a good way in demonstrating the ecologically-friendly dimensions of Buddhism (see Hanh 1990). It accentuates the commonly-shared view among Buddhists that theirs is a religion which considers environmentalism as situated at its core. This features prominently in the pronouncements of Buddhist notables such as the Dalai Lama. There are specific concepts within Buddhist doctrine that demonstrate Buddhism's irrefutable concern for the natural environment; Two of the leading of these that are most relevant here are *Paticca-samuppada,* and *Metta* (Clippard 2011).

Paticca-samuppāda This term appears with frequency in eco-Buddhist discourse. It literally means 'dependent origination;' however, it connotes 'interdependence.' This concept is of essence in any meaningful discussion of environmental stewardship as it recognizes the inextricably interconnected nature of all creatures in the universe. The acknowledgement of this truism by Buddhist is often invoked to lend credence to Buddhism's proclaimed identity as an eco-conscious religion. It is also the basis of Buddhists' critique of the dominant anthropocentric worldview. Within the framework of this thinking, humans are like other creatures, occupants of this universe. Therefore, humans co-exist and do not have dominion over other creatures as anthropocentric pronouncements would lead us to believe. In fact, Buddhists incriminate the anthropocentric worldview as a leading cause of environmental abuse. They see the concept of 'dependent origination' or 'mutual dependence' at the heart of Buddhist environmental philosophy as holding enormous promise for efforts to re-orient human understanding towards a more environment-friendly worldview. Such a worldview is invariably cognizant of the inextricably intertwined and interdependent nature of all entities in the universe. Thus, environmental stewardship should not, and does not, have as its sole objective, environmental protection. Rather, its aim must be viewed as including self-preservation—that is, the preservation of human beings. Meaningful efforts to guarantee human existence, must therefore begin with initiatives to protect the natural environment.

Mettā This concept, which connotes love, kindness, and compassion, is at the heart of Buddhist environmental ethic. Thus, Buddhism places much premium on humans loving themselves and having compassion for all living entities with which they co-habit the Earth. More importantly, the concept of *Mettā* has been linked to the Buddhist's doctrine of 'non-harm' (Clippard 2011). Of concern here is harm to all living beings. The doctrine beseeches Buddhists to develop a relationship with these beings that precludes any adversarial or harmful actions towards them. Seen from this perspective, acts such as timber exploitation for commercial purposes is frowned at. *Dhamma* or nature, therefore, constitutes or ought to constitute an integral part of built space, which is mainly the habitat of humans on a more encompassing Earth that serve as home to an assortment of other natural creatures.

5.3 Forests and Greenery in Built Space

Biodiversity loss is a nagging problem in growing cities such as those of the Asia-Pacific region. Therefore, authorities must take deliberate steps to ensure the creation and/or conservation of nature in built space. Some analysts have already observed that such steps are ongoing in cities that are experiencing post-industrial transformation (see e.g., Kowarik 2018). Thus, the natural environment is no longer considered antithetical to urban development. Instead, nature or what Kowarik

(2018: 336) has called the 'urban wilderness' "is high on the urban agenda." Such prioritization of nature in urban areas or the built environment in general, is in response to a number of growing challenges. Paramount among these is the conservation of biodiversity in demographically and physically expanding urban areas. The Food and Agriculture Organization (FAO) recognizes this and has been at the forefront of initiatives to promote urban and peri-urban forestry. As part of these initiatives, the organization coordinates and supports global, regional and sub-regional networks for the implementation of urban and peri-urban forestry throughout the world. The Asia-Pacific network held its first meeting, the 1st Asia-Pacific Urban Forestry Meeting (1st APUFM) in Zhunai, China from the 6 to 8 April 2016. The meeting brought together more than 200 participants from 17 Asian-Pacific countries. Their aim was to exchange information and experiences in managing urban and peri-urban forests in the Asia-Pacific region. The meeting's deliberations culminated in "The Zhunai Declaration of 8 April, 2016." The declaration underscored and affirmed the network's belief that "forests and trees in and around cities are the key element to make cities in the Asia-Pacific region greener, healthier, and more resilient to climate change."

The declaration is premised on the following empirically-grounded statistics. First, significantly more than half (54%) of the population of the world live in urban areas in 2014; and by 2050 more than two-thirds (66%) of the global population is projected to be urban-based. Second, Asia, a dominant part of the Asia-Pacific region contains 53% of the global urban population; it is projected to experience one of the fastest rates of urbanization within the next two decades. Third, the rapid pace of urbanization is exerting enormous pressure on the provisioning of basic services such as water and sanitation. The loss of forest and vegetation is aggravating this problem by causing severe water shortage. Urban and peri-urban forestry projects stand to contribute to efforts to reverse this phenomenon. Fourth, the Asia-Pacific region has been experiencing rapid rates of deforestation. This amplifies the need for afforestation programs such as urban and peri-urban forestry. Fifth, urban and peri-urban forestry programs can be a viable strategy for cities in the Asia-Pacific region to achieve sustainable development goals, especially Goal 11 (SDG-11). This is the goal of promoting 'sustainable cities and communities." It is impossible to exaggerate the viability of urban and peri-urban forests in efforts to achieve sustainable urban development as well as contribute to human health and well-being in urban areas.

The foregoing narrative gives the erroneous impression that urban and peri-urban forestry is new in the Asia-Pacific region. Therefore, it is necessary to state that the region has a rich history of promoting green areas in built space. This, as an FAO working paper noted, is despite the fact that significant parts of the region such as West and Central Asia are typically not located in forested areas (Akerlund et al. 2006). The West and Central Asia region, also known as commonwealth independent states (CIS) or Former Soviet Republics, constitute a sub-section of the Asia-Pacific region. There is an absence of credible statistics on the extent to which cities in this sub-region meet the World Health Organization's recommended minimum green space of 12 m^2 per capita. However, a cursory read of the relevant literature reveals that authorities in the sub-region have been committed to the creation of functioning green systems for many decades. Soviet urban planners in the sub-region have traditionally made green resources integral components of residential areas. In addition, they also made ample allowance for parks, gardens and trees along streets and roads. Furthermore, it was customary for these planners to include greenbelts to circumscribe cities for sanitary reasons; they also serve to reduce pollution from industries, protect water reserves, provide recreational areas and create a buffer between urban and rural areas (Akerlund et al. 2006). In practice, greenbelts typically comprised forests and woodlands, although in some cases, trees were planted in order to achieve a greenbelt.

To better understand the green areas in built space, or more specifically, urban green resources in the sub-region, it helps to begin by appreciating some of the major meanings that have been attributed to urban forests in the sub-region (Akerlund et al. 2006). According to Article 1 of the Forest Code of Kyrgyz Republic, a "city forest" is a forest located within a city or town. The Republic of Kazakhstan attributes a longer definition as follows (Article 4 of the Forest Code). "Urban forests and forest parks" include all forests located within an urban settlement's boundaries, registered in the State forest fund, and set aside to be used exclusively for sanitary and recreational purposes. The concept of municipal forests for authorities in Cyprus is articulated in Article 1 of the country's Forest Regulation of 1967–1991. It reads as follows.

> [A] "municipal forest means a minor state forest assigned by the Council of Ministers to a municipality for the purpose of obtaining fuel, timber and other forest produce therefrom or in order to be used and enjoyed by the citizens for the purpose of their amenities and recreation."

The authorities in Georgia have also attributed an insightful meaning to the concept of urban forests. According to Article 21 of the country's Forest Code, forested or green areas adjacent to cities and other human settlements are assigned the title green zone forests.

One typology for the main types of urban forest in the sub-regions contains five main categories (Akerlund et al. 2006). These include the *gorodskoj les, lesopark,* public green space, green space of limited public access, and special green space. The *gorodskoj les* include urban forests that are located in the urban fringe of cities or other human settlements. *Lesoparks* comprise forest parks that are located in areas at the peripheries of urban areas but contain a few pieces of recreational fixtures such as swings and sea-saws. The public green space includes parks, squares, boulevards, gardens in residential areas and micro-districts. Green space of limited public access includes green space located around private homes, nursery centres, schools, public buildings, sports complexes and industrial parks. Finally, special green space includes protective plantations that may be located along roads, nurseries and botanical gardens (Box 5.1).

Box 5.1: Urban and Peri-urban Forestry and Greening in CIS Countries

Afghanistan

Urban and peri-urban forestry and greening was about to take off in Kabul, but was abruptly interrupted by the U.S.-supported war against the Taliban. However, modest tree-planting by Kabul municipal authorities with financial support from international aid organizations have been ongoing since the early 2000s.

Armenia

Most of this country's urban forestry and greening initiatives have been concentrated in Yerevan, the national capital. This city has a reputation as one of the greenest cities in the defunct Soviet Union. It boasts lush green gardens, parks, trees planted to flank streets and boulevards, greenbelts circumscribing cities and other human settlements. However, there has been a decline in the inventory of this green infrastructure thanks to the energy crisis of the early-1990s. By some accounts, as much as 50% of the inventory has disappeared (Akerlund et al. 2006).

Azerbaijan

This country has a Forest Code and Land Legislation that is premised on the belief that forests are critical for the promotion of sanitation and public health. In addition, the code/legislation recognizes the touristic value of forest given the empirical evidence demonstrating that some visitors to the country are attracted by its rich inventory of forests. However, some parts of the country are characterized by arid soil, which is not conducive to the growth of vegetation. Despite this, authorities in the country have always been actively involved in tree-planting initiatives. During the oil boom, oil business entrepreneurs collaborated with municipal authorities in efforts to green major cities throughout the country. These efforts notwithstanding, success was sporadic because land in and around major cities such as Baku has suffered from serious oil pollution. Consequently, enormous soil enrichment efforts were necessary to permit even the most modest fertility for forest growth. This explains the fact that the only flourishing urban forests in the country are at considerable distances beyond the city limit. For instance, the urban forest for the City of Baku, which was a small medieval town before the discovery of oil there, is situated 200 km from the city.

Georgia

To appreciate the experience of Georgia with tree planting in urban centres in particular, and urban forestry in general, it helps to focus on the country's largest city, Tbilisi. Like other cities of the erstwhile Soviet Union, Tbilisi

(continued)

Box 5.1 (continued)

boasted a rich inventory of urban forests. However, the country's energy crisis of 1994–1995 resulted in a significant loss of trees. By some accounts, the city lost as many as 30,000 trees during this period (Akerlund et al. 2006). Active efforts to replenish the forest stock, or what authorities characterize as the greening of Tbilisi, had since started. These efforts have included the planting of artificial pine crops although no allowance has been made to cover the cost of managing the city's green zone. Similarly, there are no policies specifically designed to develop urban green infrastructure in Georgia as a whole and Tbilisi in particular.

Kazakhstan

Cities in Kazakhstan suffered from shrinkage or de-urbanization in the 1990s and early-2000s. This was a consequence of the dissolution of the Soviet Union in 1991 and the subsequent exodus of Slavs and Germans. In 1997 the national capital was transferred from Almaty to Astana, where it has remained since then. Although Almaty is no longer the national capital, it constitutes the ideal focus of any meaningful analysis of urban and peri-urban forestry in the country. This is because it was the leading beneficiary of urban development projects in Kazakhstan as a political unit within the Soviet Union. The urban forests, parks and green areas in the city were designed in keeping with the Soviet tradition for such facilities. Thus, the city boasts parks that serve different purposes such as cultural preservation, recreation, immortalization of memories as well as parks for the promotion of science such as botanical gardens and zoos. Furthermore, the city also contains an array of green networks that link the parks and gardens in the city centre through green corridors and boulevards to green spaces in the residential areas. The trees and other vegetation comprising this system have suffered some significant losses over the years, especially in the 2000s due to problems with the irrigation facilities.

Kyrgyzstan

Kyrgyzstan counts among the least urbanized countries in Central Asia. Its capital city, Bishkek, has a population of a little under one million (976,734 in 2017). Urban greening and forestry initiatives have been ongoing in the city since the nineteenth century when it was undergoing its initial development phase. Today, a significant portion of the city is covered by forests, none of which is private property. About two-thirds of the forests, including the land

(continued)

Box 5.1 (continued)

upon which they sit, belongs to local municipalities, while the rest is owned by the national government.

Tajikistan

Tajikistan's capital city, Dushanbe, has a population of a little more than 800,000 according to 2018 estimates. It is richly endowed with urban forests and greenery mainly in the form of green plantations that are located within housing estates and along major streets throughout the city. This are complemented by a few industrial green zones. Suburban green areas are non-existent despite their necessity given the city's hot climate.

Turkmenistan

Turkmenistan was also part of the Soviet Union until the union was dissolved in 1991. Its capital city, Ashgabat, is generously furnished with green space, including well adorned squares, complete with water fountains, trees and shrubs. The national government places a lot of emphasis on establishing greenbelts around the major cities, of which Ashgabat is the largest. Since 1998, a series of national decrees have been promulgated to stipulate the number of trees that must be planted in each city. More than 50 million tree seedlings have been planted as an upshot of initiatives on this front. On its part, the country's Ministry of Forestry has, since 1999, created an annual forestry plan. This plan is designed to produce and sell seedlings for afforestation projects in cities throughout the country.

Uzbekistan

With a population of 32,768,725 (2018 estimates), Uzbekistan is Central Asia's most populated country. Its national capital city, Tashkent, boasts a population of 2,393,000 (2018 estimates), and has been characterized as one of the 'greenest cities' in the Central Asia sub-region. Historically, the city has gone under the nickname, city-garden. In the early-to-mid-2000s, city parks, public gardens, avenues and street plantations occupied as much as 20 km^2 of the city. In addition, planted territories of dwelling districts, kindergartens, schools, hospitals, and other establishments occupied 70 km^2. During the same period, the city's green space distribution was an average of 2.5 m^2 per urban resident. Its stock of green space amounted to 24,830 ha of Municipal Gardens and Forest Parks, with more than 22,000 ha of stands of forests around cities and other human settlements, and industrial districts.

(Source: Akerlund et al. 2006).

5.4 Agriculture and Food Production in Built Space

Most cities in Asia have neither devoted the necessary attention nor enacted well thought-out policies to ensure their food self-sufficiency. Yet, the countries in this region are among the most rapidly urbanizing in the world. Urban planners in the region have been preoccupied with modernist projects. These projects often tend to emphasize more orthodox objectives of urban planning such as the compartmentalization of land use activities. Often missing from the urban planning agenda is the stipulation of steps to address questions of sustainability. Sarker et al. (2019) drew on the case of Australia to highlight this oversight in contemporary urban planning initiatives. Australia is one of the fastest urbanizing countries in the world. Therefore, concern with food sufficiency ought to be at the top of local planning agendas throughout the country. This is because food security impacts development as well as other national priorities such as public health, social equality and employment generation. More importantly, no meaningful discussion of sustainability can be deemed complete without mention of the question of food self-sufficiency. To be sure, urban planners and authorities in Australia and the Asia and Pacific region recognize the importance of urban agriculture as a viable strategy to guarantee food self-sufficiency.

Food self-sufficiency is generally not accorded the attention it deserves on the urban planning agenda in the region. However, the conduct of municipal authorities in Australia is indicative of a strong commitment to pursuing this objective. The City of Sydney, for instance, has operated a network, the Sydney Peri-Urban Network (SPUN) dedicated to promoting peri-urban agriculture since 2013. The network comprises 12 councils that border metropolitan Sydney. These include, Blue Mountain, Camden, Gosford, Hawkesbury, Hornsby, Kiama, Penrith, Shellharbour, Shoalhaven, The Hills Shire, Wingecarribee, and Willondilly. Its avowed aim is to ensure local food self-sufficiency. The avowed vision of SPUN can be re-stated to include the following: assisting and empowering councils to manage the significant pressures they face in peri-urban areas. The premise for this is the general belief by SPUN in the ability of peri-urban agriculture to strengthen its broader local area, New South Wales. In particular, it believes that peri-urban agriculture can boosts New South Wale's attractiveness as a touristic destination, and its character as a place to call home. Among the many roles that SPUN has committed to undertake are the following: advocate for the recognition of urban and peri-urban areas across New South Wales (NSW); serve as a platform for the dissemination of information and knowledge on good practice; build partnership and share resources in the peri-urban farming domain; promote connections and productive cross-fertilization; develop the economic potential of peri-urban areas; ensure the long term viability of peri-urban agricultural activities; achieve the sustainability of managed open space with a view to promoting biodiversity; improve multi-functional land use; and manage land use change within peri-urban areas.

Although long, the foregoing is not an exhaustive list of the many ways in which urban and peri-urban agricultural bodies in Australia are contributing to environmental stewardship. As the narrative in Box 5.2 suggests, such bodies have been known to

Box 5.2: Perth City Farm, East Perth, Australia
Launched in 1994, Perth City Farm (PCF) is a non-profit facility located between the railway and the Education Department in East Perth, Australia. One of its most notable features for environmental stewardship purposes is the fact that it sits on a piece of reclaimed land previously occupied and environmentally polluted by a scrap metal, battery and train recycling warehouses. When Perth City Farm (PCF) first assumed possession of this 7600 m^2 lot and the three warehouses occupying it, the land was terribly contaminated by years of dumping toxic fluids and scrap metal. City Farm, therefore, undertook to detoxicate and make the land suitable for farming. Detoxification entailed eliminating heavy metals, hydrocarbons, and acid from the land. In addition, it was necessary to remove the top soil up to a depth of about 60 cm to 1 m, and replacing this with compost manured soil. Currently, Perth City Farm is a leader in urban and peri-urban farming in Australia. Its team also runs two major sites in the Perth Central Business District (CBD). One is the Urban Orchard and Wetlands at the Perth Cultural Centre, and the other is the QV1 Rooftop Garden. The PCF also created an Inner City Greening Team charged with the responsibility of building connections between people and the natural environment in Perth City. Another notable activity of the PCF is the promotion of environmental awareness. In this regard it organizes environmental training programs from time to time. In 1994, for instance, it organized a 26-week Land care and Environment Action Program (LEAP) training course. The course was designed to teach participants nursery and land conservation techniques, plant propagation, rural and urban land care, permaculture design, and horticulture. The participants were mainly unemployed youth between the ages of 15 and 25. Administered by the Western Australian Department of Training and Men of Trees, Western Australia, the training was sponsored by the (Australian) Federal Department of Employment, Education and Training.

Source: PCF (Online).

play important environmental stewardship roles such as land reclamation and detoxification, cultural preservation and the promotion of environmental awareness. The narrative also demonstrates the ability of urban and peri-urban initiatives to create a healthy work-life balance within the community. Thus, lessons of experience from Australia suggest that urban and peri-urban agriculture does more than just increase the food supply of urban inhabitants. It improves people's quality of life not only at home but at work. At home, people transform small areas such as their backyards, front porch and portions of their lawns to grow vegetables. At work, urban gardening affords workers the opportunity to break the monotony of work to create and care for corporate gardens. Here, as the narrative shows, workers have been known to grow fruits, herbs and vegetables for office meals; they have also been known to cultivate flowers to decorate office space or use at private social events (PCF Online). Finally, the narrative underscores the importance of government support in ensuring the success of urban and peri-urban gardening and cognate initiatives.

Australia is among the few countries that prioritize urban and peri-urban agriculture as an issue of regional importance in the Asia and Pacific region. Thus, despite their importance, urban and peri-urban agriculture are treated literally as fringe activities. Three reasons may account for this. The first is the tendency to relegate issues related to food production to a tertiary position unless there is a famine. Therefore, as long as there is enough food for everyone, policy makers typically never consider food production an item deserving of a place on the national or municipal policy agenda. The second is the seemingly natural inverse relationship between economic growth and food production within any given geographic space. Thus, as discussed below, economic prosperity in some of the Asia and Pacific countries has resulted in a significant decline in food production within those countries.

Finally, there is the sheer lack of foresight on the part of policy makers in some countries that are experiencing rapid rates of urbanization that are not necessarily matched by corresponding levels of economic growth. Yet, the need for urban and peri-urban agriculture is accentuated rather than diminished in such countries. Generally speaking, the factors inhibiting urban and peri-urban agriculture in the Asia and Pacific region are plentiful. However, they can be summarized under five categories as follows: urban growth and sprawl, loss of fertile soil in areas within and proximate to cities, inappropriate or antiquated urban planning legislation and laws, lack of information systems to collect and process data, and economic progress.

Urban Growth and Sprawl The hyper-urbanization trends that have been experienced by Asia-Pacific countries have resulted in the rapid loss of land at the peripheries of urban centres. These are areas that would ideally serve as sites for urban and peri-urban agriculture. Some analysts have drawn attention to the crucial role that such sites served in efforts to meet the food needs of large cities in China half a century or so ago (see e.g. Yeung 1987). For instance, Taipei, China, depended on such sites for 70% of its vegetable supply in the 1960s. By the early-1970s, this proportion had dropped to a mere 30%. Beijing also experienced a similar transformation of its physical space during the same period. The disappearance or severe diminution of agricultural land resulted from many factors with the reduction of agricultural holdings in the face of urban expansion being the most prominent. South Korea recorded a similar trend in which urban expansion resulted in a significant decline in agricultural land in and around major cities. More specifically, some 1066 km^2 of agricultural land was converted to non-agricultural uses in the 1970s and 1980s. The common denominator in the Chinese and South Korean cases is economic prosperity. Economic prosperity in both countries led to the enactment of policies that were primarily market-oriented at the expense of family-oriented and subsistence farming alternatives. It also fueled large investments in construction projects that gobbled up most of the prime land within cities and their outskirts.

Paradoxically, the loss of arable land at the urban fringes resulted in freeing up patches of land dotted among buildings in residential, commercial and industrial districts, around parking lots and along highways and under overhead high tension electric cable lines. However, because of the built-up nature and ownership issues surrounding land at the heart of cities, its utility as farmland has never been maximized. Consequently, the loss of arable land at the urban fringes has translated into the availability of vacant but highly under-utilized land in urban centres. Another problem with urban agriculture in the region has to do with town and country planning laws and regulations. These are either antiquated or simply not apropos, or both. Given the influence of Europeans and other agents of Western civilization, there is a tendency to uncritically favour modernist urban planning principles. These are best known for their penchant for orderliness entailing the segregation of land use activities. Under the stipulations of modernist planning, living and residential activities in general are considered incompatible with agriculture. In fact, agriculture is viewed as an undesirable land use activity within the urban milieu.

The economic prosperity recorded by many countries in the region during the last couple of decades has also contributed to the loss suffered by urban agriculture. In this case, economic prosperity has led to increased dependence on imported food in countries experiencing significant economic growth. Countries exemplifying this trend in the region include, but are not limited to Singapore, Japan, China, South Korea, and Malaysia. Thus, not all countries fall under this category. Rather, quite a good number of countries in the region have not experienced the level of economic prosperity enjoyed by the afore-mentioned. Countries in this latter group, prominent among which are Bangladesh, India, Pakistan, Vietnam, Cambodia and Thailand, continue to locally produce almost all of the food they consume.

Nevertheless, a number of urban and peri-urban agricultural activities are observable in the region. At the forefront of these are an increase in agricultural activities per unit of space, low capital input per unit of production, low energy consumption, low marketing cost and freshness of products. Although urban agricultural activities in the region vary significantly among countries, a few commonalities are identifiable. These include, the dominance of fish farming, the increasing trend in favor of cash crops and livestock at the expense of food grain production. The rarity of urban forestry meant to increase food supply has not boded well for efforts to attain food self-sufficiency within countries in the region. Yet, it should not be lost on any meaningful analysis of food production initiatives in built space that urban and peri-urban agriculture, especially in the form of urban gardening, constitutes a significant element of the region's history.

Given the resilience of fish-farming as an agricultural activity in built space in the region, it is order to discuss it in greater detail. Fish has always been the source of animal protein intake for people in the Asia-Pacific region from time immemorial. Currently, it is the source of 50% of the animal protein intake for most of the region. This percentage is significantly greater in some countries. For example, fish comprises about 70% of animal protein intake in South Korea, and Indonesia. While most of the fish eaten in the region comes from the sea and freshwater, a considerable portion originates in aquacultural sources, especially fish ponds. In fact, this is the source of half (50%) of the fish consumed in China 9.2% of the fish consumed in Indonesia, and 10.1% of that consumed in the Philippines. Aquacultural activities grew by 50% in these countries in the 1970s.

An examination of urban and peri-urban agricultural activities in Vietnam can prove to be very informative. This type of agriculture has a long history in Vietnam. In fact, urban agriculture has been practiced in this part of the world long before it became a popular undertaking within the international development community. Today, major Vietnamese cities such as Hanoi, the national capital, and Ho Chi Minh City or what French colonial authorities called Saigon, are richly punctuated by urban farms and/or gardens. These are typically tucked between buildings, along major thoroughfares, and around public buildings. In addition, the outskirts of these cities are buzzling with peri-urban agricultural activities. Yet, urban agriculture is not without its own share of problems. Tinh et al. (2007) have identified some of the constraints confronting urban agriculture in the country. Prominent among the contraints are: the quality of urban agricultural products, supply seasonality of local markets, lack of the capacity to handle the negative externalities of noxious urban agricultural activities. Examples of such activities include pig rearing, and poultry. Also, they are environmental pollution concerns arising especially from the use of fertilizers. While chemical fertilizers and pesticides are important, and sometimes required for crop growth and survival they are occasionally over-used. This often results in environmental pollution. In addition, the quality and safety of urban farm products, particularly vegetable in some parts of the country. For example, vegetables from two rural districts were found to have been contaminated in the early-2000s.

5.5 Energy Production and Consumption in Built Space

A substantial proportion of the households without access to clean energy in the world is based in the Asia and Pacific region. One estimate reveals that as many as 42% of those without electricity, for instance, live in South Asia alone (Palit and Chaurey 2011: 267). By some accounts, the number of those without access to electricity in the region is at least half a billion (ADB Online). The United Nations Economic Commission for the Asia and Pacific region (UNESCAP 2018) puts this figure at 420 million. This is about half of the entire region's population. The problem of clean energy deprivation promises to grow rather than decline within the foreseeable future. According to one projection, the demand for energy will almost double in 2030 (ADB Online, para 1). These statistics can be best understood by also taking into account the tremendous progress registered by efforts to improve access to electricity in the region during the last two or so decades.

These efforts are unlikely to succeed unless they incorporate initiatives to promote renewable production and consumption. Presently, and according to the relevant statistics provided by the United Nations Economic Commission for Asia and the Pacific (UNESCAP 2018), efforts in this connection leave much to be desired. Particularly, the rate of renewable consumption remains low. In 2014, this rate was estimated to be 18.3% (UNESCAP 2018). This estimate includes renewable energy from both traditional and modern sources such as hydro, solar, wind, geothermal, and modern biofuels. The 2014 rate of 18.3% actually constituted a decrease from the 1990 rate of 23%, but an increase from the 2011 rate of 17.9%. Another important statistic from 2014 worth mentioning here relates to the proportion of total final energy consumed that was of the renewable genre. This proportion stood at 6.8%, which represented an increase over the 2012 proportion of 6.2%. If everything remains on the current course, the region will attain the Sustainable Development Goal's (SDG) target of universal access to energy. This is Goal 7 of the SDGs. This projection belies the fact that many countries in the region, especially those in the Pacific, remain without access, or have very limited access to improved energy.

A number of other statistics contained in the UNESCAP report are also very telling. For instance, between 2012 and 2014, 93.1 million persons in the region gained access to electricity. This was a significant increase over the 2012 rate of 83.3 million. The region's urban areas are steadily progressing towards universal access with respect to electricity. In 2014, the electricity access level stood at 98.7%. However, rural areas continue to lag behind with only 83.3% reported as having access to electricity in 2012. From 2012 to 2014, China, India and Pakistan each added between 13 and 16 million persons to their respective populations. Afghanistan, Bangladesh, Indonesia, and Philippines extended electricity services to between five and nine million people. Some 2.1 billion people in the region, that is, about half of the region's total population, or a quarter of the global population, remains without access to electricity. This population depends on unsafe energy sources, which are harmful to humans. In this regard, the World Health Organization (WHO) estimates that as many as 92 deaths per 100,000 due to household air pollution in developing Asian nations. In 2014, the region recorded an access rate to clean cooking fuel and technology of 51.2%. This represents a tremendous increase over the 2000 figure of almost 40% (39.8%). Only 12 countries, including Australia, Brunei Darussalam, Japan, New Zealand, Singapore, South Korea, Malaysia, Turkmenistan, Russia, Iran, Maldives, and Azerbaijan in the region have full coverage with respect to clean cooking technology (United Nations ESCAP 2018: 32).

These statistics suggest that the goal of attaining full coverage or universal access to clean cooking technology by 2030 remains elusive. Here, it is worth noting that the average annual share of increase in access to this technology was around 0.8% between 2000 and 2014. This falls well below the pace to achieve the projected coverage (UNESCAP 2018: 3). Some countries in the region constitute the exception. Examples include China and Australia. These are in fact, the regional leaders in renewable energy adoption initiatives. China's ambitious renewable installation program is committed to investing as much as $360 billion in the renewable energy sector by 2020. As for Australia, the government of that country has taken active steps to promote renewable energy. Thus far, it completed and launched 700 MW of renewable energy projects in 2017 alone. At the same time, investments in large wind and solar projects and related activities in the country rose by 150% to a record $9 billion. The authorities in Australia have set a goal to source 23.5% of the country's electricity from renewables sources by 2020. These authorities are actually on track to exceed this target. However, the same does not hold true for the rest of the region. As stated earlier, the region's record on promoting access to clean energy sources leaves something to be desired. Therefore, there is an urgent need to craft sagacious strategies to meet the region's energy demand within the foreseeable future. To be considered sagacious, the strategy must be socially, economically and environmentally sustainable.

The region's energy problematic is exacerbated by high levels of poverty in the region. The Asian Development Bank (ADB) has drawn attention to some issues deserving of immediate attention. These include, undertaking meaningful sector governance reforms, and indulging in intra-regional and international cooperation. Sector governance, the ADB argues, is necessary to secure access to energy. It is also a means of significantly improving efficiency in the energy sector, and reducing energy cost. As for intra-regional and international cooperation in the energy domain, this is crucial to maximize the utility of scarce energy resources within the region. For instance, sub-regional power trade promises to assist municipal authorities fulfill local energy demand.

5.6 Water Production and Consumption in Built Space

The Asia-Pacific Region is not only vast, it is also diverse. This diversity is especially acute with respect to freshwater endowment. In this regard, the region contains water-abundant areas in the Himalayan as well as water-stressed small island states (UNEP 2012). Nevertheless, the region as a whole faces a problem of limited access to water. Nearly 1.7 billion people in the region have no access to improved drinking water and sanitation. The problem is partially a function of the fact that water withdrawals in the region have been speedily increasing in response to demands from rapidly growing populations. However, it is necessary to state that most of the water withdrawn from groundwater sources is for agricultural purposes. Farmers in northwest India, northeast China, and northeast Pakistan depend on groundwater for irrigation. As a United Nations Environmental Programme report stated, groundwater withdrawal in the region increased by 230% between the 1980s and 2000s (UNEP 2012). The lack of access has been characterized as a silent crisis that has claimed more lives than any conflict in the region. The World Water Development Report 2015 by UNESCAP in collaboration with UNESCO, other UN-Water members and 37 partners of immense importance to the region's water situation. The report emphasizes the interconnectivity among water related disasters, urbanization and the effects of ground water irrigation. It suggests that considerable progress has been made in efforts to attain universal access to water in the region. The percentage of people without access to water fell from 27% in 1990, in 1990 to 8% in 2012. On the whole, China and India registered about half of the world's progress towards attaining universal access to water during the 2005–2015 decade (UN.Org 2019). India and China added 522 million and 457 million persons respectively to the list of those with access to improved water between 1990 and 2010. Eastern Asia, which dominated demographically and economically, recorded an impressive gain of 35% points with respect to access to improved water in two decades (1990–2010) (Xinhua 2018). Southern Asia is also on record for its remarkable progress on this front. The sub-region registered a gain of 65% points in access to water between 1990 and 2010

(Xinhua 2018). Here, improved access is operationalized at two levels. The first groups together people with access to, or use, piped water. This group is considerably small and includes only 30% of the total with access to improved water. The countries in this group are mainly located in three sub-regions, including Oceania, Southern Asia and South-Eastern Asia. The second category contains persons with access to piped water on their premises. This category includes 70% of the population, located mainly in the Eastern and Western Asia subregions.

However, these gains must not detract from the fact that the built environment in significant portions of the region remain water-deficient. In particular, millions of people in the region remain without access to improved water. By some estimates, as many as 300 million people in the Asia and Pacific region lack access to improved water sources (Xinhua 2018). Projections paint a bleak picture for the future as some 3.4 billion people in the region are likely to face water insecurity and related problems by 2050 (Xinhua 2018). More worthy of note is the fact that the gains with respect to access to improved water have not been universal throughout the region. Some subregions, especially the Caucasus, Central Asia, and the Oceania subregions have actually regressed. The reasons for this are plentiful with the lack of financial resources constituting the most dominant. This problem is a function of a rapidly shrinking tax base in countries of the defunct Soviet Union, which have witnessed a heavy population decline during the last couple of decades. Thus, many countries in the region continue to face problems attaining universal coverage with respect to access to improved water. Paradoxically, this problem is just as serious in countries experiencing rapid population growth and hyper-urbanization as it is in those suffering from population shrinkage.

The problem is heightened in countries that depend mainly on groundwater sources; while the sector most impacted is agriculture. Accordingly, countries that depend on groundwater for irrigation are likely to experience serious water stress and related problems within the foreseeable future unless meaningful steps are taken to mitigate the problem. Five countries, including Bangladesh, China, India, Nepal, and Pakistan are particularly worth noting in this regard. Together, these countries use as much as half of the world's total groundwater (UNESCO 2015). In general, these problems tend to be exacerbated by extreme weather events of which many are likely to occur in the region within the foreseeable future. Prominent in this connection are climate-change-related catastrophes, which are projected to intensify the region's water stress difficulties (UN-Water 2019). The following also feature prominently among the problems likely to result from climate and concomitant eventualities: strong river flow variations, rainfall variability and increased occurrence of flood and drought events; droughts and related extreme weather events.

The water problems of the Asia and Pacific region are not entirely of the natural variant; some are of the institutional and technological genres. The institutional framework for water service exploitation and distribution in the region typically assumes two forms, viz., fragmented (as in the case of Thailand 5.4) and overcentralized (as in the case of China). In either case, authorities have not demonstrated a propensity for enlisting the participation of service beneficiaries. Rather, there has been a tendency to outsource water service delivery responsibilities to private corporations (Box 5.3). This is especially true in large cities of countries

Box 5.3: The Water Situation in Thailand

Thailand faces challenges in five specific water-related areas. These include shortage of drinking water for domestic use in rural milieus; water for irrigation; water for rapidly expanding urban areas; water for industrial development; and water for ecosystems and pollution control. Government efforts to address these problems have been multifaceted, but mainly targeted at integrating an ostensibly fragmented institutional framework. The main plan in this regard reposes on five main pillars, namely the enlistment of different agencies in related policy fields to constitute an interorganizational framework for water policy implementation; the integration of forest, land and water resources, a refreshed administrative approach incorporating river basin interests including upstream, midstream, and downstream users; attentiveness to budgetary issues; and the summoning of community participatory strategies entailing the incorporation of stakeholders and their interests at all levels. The plan contains five different but overlapping sections dealing respectively with: rehabilitation of upstream areas, initiatives to address water shortage issues; protection and alleviation of food disasters; efforts to address water quality problems (e.g., through water treatment projects); and the promotion of water management efficiency, for example, by use of institutional and legal instruments (e.g., water law). As part of the efforts to promote water management in the country, a number of river commissions have been constituted. One of these, the Mekong River Commission stands out. It is charged with, among other things, promoting awareness of the Bung Khong Long Wetland through networking, stakeholder regulation and participation and social sanction mechanisms. This was designed to promote environmental restoration and reforestation. These initiatives have one primary goal, namely the promotion of sustainable water exploitation and use.

A specific case of efforts focused mainly on such promotion can be found in the Mea Moh District, Lampang Province. This case is illustrative of the inextricable link connecting food, energy and water. It involves The Electricity Generating Authority of Thailand (EGAT), which was established more than three decades ago. The corporation operated 13 stream power plants that employ lignite as fuel to generate electricity. The plants and the entire electricity generation process has been blamed for causing air and water pollution. These have negatively impacted health conditions for populations in the proximate area of the plant. To deal with this problem, a project was created under the name, 'System and Mechanism Development By Participation of Community Near-By Mae Moh Plant in Water Supply Provision For Agriculture And Consumption.' The project's main charge is to provide safe drinking water access to rural areas. To meet this objective, action was taken to facilitate access to water in rural areas in the upper zone of Ban Dong Sub-District within the Mea Moh District in the Lampang Province of Northern Thailand. Source: UN-Water (Online).

such as the Philippines, Indonesia and Malaysia, where long-term concessions have always dominated the private water utility market (Bluefield Research 2019). Malaysia witnessed a change in this respect in 2015. This is when the Malaysian government moved to nationalize the country's utility sector. Nevertheless, private utility corporations continue to dominate the water sector; in 2015 these corporations supplied as many as 3.4 billion cubic meters of drinking water to more than 40 million people in South-East Asia alone (Bluefield Research 2019). The water problematic of the region also has technological dimensions mainly having to do with recycling used water. The statistics in this regard leave much to be desired; in particular, only a very small proportion of used water in the region is recycled. Consider the following examples: the percentage of recycled wastewater is only 14 in Indonesia, 9 in India, 10 in the Philippines; and 4 in Vietnam (Galang 2016). The problem is rendered more complicated by the fact that some of the important sources of freshwater in the region straddle international boundaries.

Thus, the problem is of a transboundary nature. This means, among other things that access to freshwater resources constitute a viable source of international conflict in the region. To be sure, the region is no stranger to such conflicts. The following three examples are of recent vintage. The first is the conflict between Malaysia and Singapore arising from the former's constant threat to cut off water supply to the latter. The second involves Bangladesh and India. Here, Bangladesh has, over the decades, been accusing her domineering neighbor, India, for limiting the quantity of water flowing to her from the Ganges River. The third is the conflict between China and her significantly smaller neighbors of the Mekong Delta. These neighbors have tried unsuccessfully to have China participate in multinational agreements that would assure them unbridled access to water from the Mekong River.

Despite China's privileged access to the Mekong River, her water problems still require careful attention. This is not lost on Chinese authorities; they have been considerably proactive by crafting innovative water rationing policies. Among these are the systems of water use quotas and a set of fee collection rules. These were designed to address the severe water flow problems associated with harvesting water from the Yellow River Basin.

5.7 Conclusion

The Asia and Pacific region is diverse in many ways; in particular, it contains countries of different sizes, socio-economic development levels, and level of natural resource endowment. At the same time, the countries share much in common. For one thing, the countries are deeply religious. For another, they have been experiencing several problems, many of which are caused by levels of urbanization, population growth, and industrialization. These factors have engendered much needed economic growth. However, the factors have also been found at the root of problems that obfuscate efforts to promote nature in the built environment such as air, noise, traffic congestion, and water pollution. Major cities in the region, especially Shanghai in China, Tokyo in Japan, Jakarta in Indonesia and New Delhi in India are already experiencing intolerable levels of air and noise pollution. This chapter contends that to understand the challenges faced by efforts to promote nature in built space in the region, it helps to first appreciate its politico-economic, socio-cultural, geo-ecological and historical contexts. The chapter has shed light on the functioning of these contexts. In particular, the chapter has discussed the various ways by which political, economic, social, technological, ecological, culture and historical factors affect initiatives to create and maintain forests, gardens, parks, and green infrastructure in general in the built environment in the Asia and Pacific region.

References

Akerlund U, Knuth L, Randrup TB, Schipperijn J (2006) Urban and Peri-Urban Forestry and Greening in West and Central Asia: experiences, constraints and prospects. LSP working paper 36. Food and Agriculture Organization (FAO), Access to Natural Resources Sub-Programme

Badiner AH (ed) (1990) Dharma Gaia: a harvest of essays in Buddhism and ecology. Parallax Press, Berkeley, CA

BBC (Online) What does Hinduism teach about the environment? Accessed 10 Sept 2019 via: bbc.co.uk/bitesize/guides/zbvrq6f/revision/3

Bluefield Research (2019) Southeast Asia private water utilities: company rankings, 2016. Accessed 23 Sept 2019 via: https://www.bluefieldresearch.com/research/southeast-asia-private-water-rankings/

Clippard S (2011) The Lorax wears saffron: toward a Buddhist environmentalism. J Buddh Ethics 18:211–48

ECOASIA (Online) A long-term perspective on environment and development in Asia-Pacific region. An E-Book by the Environmental Congress for Asia and the Pacific (ECOASIA) Available online. Accessed 8 Sept 2019 via: https://www.env.go.jp/earth/ecoasia/workshop/bluebook/chapter1-1.html

Escobedo F, Seitz J (2009) The costs of managing an urban forest. Institute of Food and Agricultural Sciences, University of Florida. Report NE-186, Radnor, PA (1994), 83–94

Galang J (2016) Asia-Pacific 'hot spot for water insecurity.' Feature article, scidev.net. Accessed 23 Sept 2019 via: scidev.net/global/water/feature/asia-pacific-hot-spot-for-water-insecurity.html

Grim JA (2019) Indigenous lifeways and ecology. Available online. Accessed 14 Sept 2019 via: http://fore.yale.edu/religion/indigenous/

Hanh TN (1990) Earth gathas. In: Badiner AH (ed) Dharma Gaia: a harvest of essays in Buddhism and ecology. Parallax Press, Berkeley, CA, p 195

Hope M, Young J (2019) Islam and ecology. Cross currents. Available online. Accessed 14 Oct 2019 via: https://crosscurrents.org/islamecology.htm

Kowarik I (2018) Urban wilderness: supply, demand, and access. Urban For Urban Green 29:336–347

Nasr SH (1968a) Man and nature: the spiritual crisis of modern man. George Allen and Unwin Ltd., London

Nasr SH (1968b) Science and civilization in Islam. Harvard University Press, Cambridge

Nasr SH (1972) Sufi essays. George Allen and Unwin, London

Nasr SH (1976) Islamic science: an illustrated study. World of Islam Festival Publishing Co. and Thorson Publishers, London

Nasr SH (1987) Traditional Islam and the modern world. Kegan Paul, London

Nasr SH (1989) Gifford lectures. State University of New York Press, Albany

Palit D, Chaurey A (2011) Off-grid rural electrification experiences from South Asia: status and best practice. Energy for sustainable development, 15, 266–276

PCF (Online) Inner city gardening. Perth City Farms (PCF). Available online. Accessed 19 Sept 2019 via: https://www.perthcityfarm.org.au/city-greening

PRC (2015) The future of world religions: population growth. Pew Research Centre (PRC). Demographic study (April 2015). Available online. Accessed 10 Sept 2010 via https://www.pewforum.org/2015/04/02/asia-pacific/

Sarker AF, Bornman JF, Marino D (2019) A framework for integrating agriculture in urban sustainability in Australia. Urban Sci 3(50). https://doi.org/10.3390/urbansci3020050

Song XP, Tan PY, Edwards P, Richards D (2018) The economic benefits and costs of trees in urban forest stewardship: a systematic review. Urban For Urban Green 29:162–170

Tinh NT, Warnaars M, Duyen TTB, Ngoc TTB (2007) Farming in the city: an annotated bibliography of urban and peri-urban agriculture in Vietnam with emphasis on Hanoi. Limi Peru: Urban Harvest Global Coordination Office, C/O. International Potato Center (CIP)

UNEP (2012) Summary for Asia and the Pacific region: On the Eve of Rio+20. Global Environmental Outlook (GEO5). Retrieved on September 25, 2019 from: http://www.unep.org/geo

UNEP (2016) The emissions gap report 2016. United Nations Environmental Programme (UNEP), Nairobi, Kenya

UNESCO (2015) Towards sustainable groundwater management in Asian cities. In: UNESCO, Facing the challenges: case studies and indicators. UNESCO, Bangkok

UNESCAP (2018) Achieving SDG7 in Asia and the Pacific. Policy Brief #19. United Nations Economic Commission for Asia Pacific (UNESCAP) in Collaboration with ADB, AIIB, FIA Foundation, IEA, IRENA and UNDP

UN-Water (2019) Special regional session: achieving water security for Asia and the Pacific through sustainable water management. Session report, 15th January 2015. Accessed 24 Sept 2019 via: https://www.un.org/waterforlifedecade/waterandsustainabledevelopment2015/pdf/sesiones/Regional_Session_Asia%20and%20Pacific.pdf

United Nations ESCAP (2018) Asia-Pacific progress in sustainable energy: a global tracking framework 2017 regional assessment report. Available online. Accessed 27 Jan 2019 via: https://www.unescap.org/sites/default/files/publications/A%20Global%20Tracking%20Framework_Web%202018_0.pdf

Xinhua (2018) 300 million people in Asia Pacific have no improved access to water: ADB. Available online. Accessed 22 Sept 2019 via: sinhuanet.com/English/2018-10/02/c_137507412.htm

Yeung Y (1987) Examples of urban agriculture in Asia. Food Nutr Bull, 9(2): 14–23. The United Nations University Press

Chapter 6
Nature in Built Space in Latin America and the Caribbean

Abstract The Latin America and Caribbean region contains countries of different sizes that have, without exception, experienced European colonial domination. Most of the countries had since gained independence. However, a few, located mainly in the Caribbean remain dependencies of European countries. Examples of these include Guadeloupe and Martinique, which are administrative divisions of France, and Aruba, a dependency of the Netherlands. Therefore, efforts to create and maintain nature in built space in these countries are, or have been, typically influenced by both endogenous and exogenous factors. While this fact may be indisputable, the exact impact of these factors remains unclear. This chapter sheds light on the nature of the factors.

6.1 Introduction

This chapter focuses on nature in built space in Latin America and the Caribbean region. This is the region south of the United States, including all the islands to the south-east (see Fig. 6.1). It includes 34 sovereign countries and 15 dependent territories (see Tables 6.1 and 6.2). The countries and territories are demographically, politically and socio-culturally diverse. Brazil, for instance, is ranked as the world's seventh largest economy. At the same time, the region is home to economically budding island states such as St. Lucia and Barbados. A common denominator for all countries in the region is European colonialism. The countries, without exception experienced the colonial control of one European power or another; some of the countries remain dependencies of European powers. In the case of francophone Caribbean island territories such as Martinique and Guadeloupe, they actually exist as administrative divisions *(départements)* of France. However, the countries of mainland central and south America are mainly erstwhile colonies of Spain. The only exception is Brazil, which fell under the colonial jurisdiction of Portugal. Spain and Portugal controlled the south and central America from the thirteenth to nineteenth century. On their part, the Caribbean islands, comprising 13 sovereign territories and 15 dependencies, came under the colonial orbit of Spain, Britain, Holland and France.

To appreciate the relationship between nature and built space in the Latin America and the Caribbean (LAC) region, it is important to understand its colonial

A. J. Njoh, *Nature in the Built Environment*,
https://doi.org/10.1007/978-3-030-39759-3_6

Fig. 6.1 Map of Latin America and the Caribbean

and pre-colonial history. More importantly, it is necessary to understand the dominant belief systems and urbanization trends in the region. This chapter sheds light on these dimensions of the region's identity and interrogates the major ways by which they affect efforts to address the food, energy and water needs in its built environment. The history of pre-Columbian Latin America—that is, Latin America before the arrival of Christopher Columbus who is credited in Eurocentric accounts with the 'discovery' of the Americas—boasts an exceedingly rich history. I delve deeper into this history below. For now, suffice to state that prior to the European conquest, the indigenous people of the region had domesticated plants and developed sophisticated strategies to co-exist with nature. However, significant levels of urbanization did not occur prior to the colonial era. In its "Global Environmental Outlook Regional Assessment for Latin American and the Caribbean" the United Nations Environmental Programme (UNEP) presents recent statistics on urbanization for the region (see UNEP 2016). In 2015, the region's population stood at 626 million, an increase of 6 million since 2009. The most urbanized major region in the world, LAC's urban population grew from 79% in 2010 to 80% in 2015. As many

Table 6.1 Latin American and Caribbean countries by population, 2019

	Country	Population (2019)	Sub-region
1.	Brazil	211,049,527	South America
2.	Mexico	127,575,	Central America
3.	Colombia	50,339.443	South America
4.	Argentina	44,780,677	South America
5.	Peru	32,510,453	South America
6.	Venezuela	28,515,829	South America
7.	Chile	18,952,038	South America
8.	Guatemala	17,581,472	Central America
9.	Ecuador	17,373,662	South America
10.	Bolivia	11,513,100	South America
11,	Cuba	11,333,483	Caribbean
12.	Haiti	10,263,077	Caribbean
13.	Dominican Republic	10,738,958	Caribbean
14.	Honduras	9,746,117	Central America
15.	Paraguay	7,044,636	South America
16.	Nicaragua	6,545,502	Central America
17.	El Salvador	6,453,553	Central America
18.	Costa Rica	5,047,561	Central America
19.	Panama	4,246,439	Central America
20.	Uruguay	3,461,734	South America
21.	Jamaica	2,948,279	Caribbean
22.	Trinidad & Tobago	1,394,973	Caribbean
23.	Guyana	782,766	South America
24.	Suriname	581,372	South America
25.	Belize	390,353	Central America
26.	Bahamas	389,482	Caribbean
27.	Barbados	287,025	Caribbean
28.	Saint Lucia	182,790	Caribbean
30.	Grenada	112,003	Caribbean
31.	St. Vincent & Grenadines	110,589	Caribbean
32.	Antigua and Barbuda	97,118	Caribbean
33.	Dominica	71,808	Caribbean
34.	St. Kitts & Nevis	52,823	Caribbean

as 15 cities in the region transitioned from the 'small' to 'medium-size' category between 2005 and 2015. The same decade also witnessed the number of mega-cities in the region increase by two.

The Latin America and the Caribbean (LAC) region is characterized by extreme socio-economic inequalities rooted in a history of unequal land distribution (Huber et al. 2006). From the colonial era until this day, the region has always been dominated by large landholders. Democratic institutions are scarce throughout the region. This, some have opined, is a function of the tendency of the elite class to

Table 6.2 Dependent territories in Latin America and the Caribbean

	Territory	Population (2019)	Controling power
1.	Puerto Rico	2,933,408	USA
2.	Guadeloupe	400,056	France
3.	Martinique	375,554	France
4.	French Guiana	290,832	France
5.	Curaçao	163,424	Netherlands
6.	Aruba	106,314	Netherlands
7.	U.S. Virgin Islands	104,578	USA
8.	Cayman Islands	64,948	UK
9.	Sint Maarten	42,388	Netherlands
10.	Turks and Caicos	38,191	U.K.
11.	British Virgin Islands	30,030	U.K.
12.	Caribbean Netherlands	25,979	Netherlands
13.	Anguilla	14,869	U.K.
14.	Montserrat	4989	U.K.
15.	Falkland Islands	3377	U.K.

discourage the development and growth of democracy (Huber et al. 2006). For a considerable duration subsequent to the demise of colonialism in the region, many countries have either been ruled by dictators or experienced political instability. On the whole, countries in the region have suffered serious economic problems.

The underlying premise of this chapter is that efforts to foster the sustainable existence of nature in built space in LAC can only succeed when they repose on a good understanding of the region's history. What is the relationship between built space and the natural environment in the region from historical and contemporary perspectives? To what extent have initiatives to promote the production and/or management of clean air (via afforestation), food, energy, and water been made in the region? The chapter seeks to address these questions. It takes off in the next section by painting a compressed picture of the history of, and belief systems in, Latin America and the Caribbean.

6.2 Historical Background

Eurocentric narratives on Latin America and the Caribbean region, gives the impression that the region was without human life prior to the so-called 'discovery of the Americas' in 1492. Yet, nothing could be further from the truth. For thousands of years prior to the arrival of Columbus and other Europeans, indigenous peoples of the region had developed sophisticated techniques and strategies to harmoniously co-exist with nature. One remarkable innovation that demonstrates an incredible ability to accommodate nature in built space is the domestication of plants. It is in this regard that, although separated by thousands of kilometers, the so-called 'New

World' (the Americas, including Latin America and the Caribbean) and the 'Old World' (comprising the Middle-East and Europe) share much in common. For instance, at about the same time as agriculture was emerging in the Old World, people in what has come to be known as Latin America and the Caribbean were growing crops. According to one authoritative account, there is a plethora of evidence indicating that squash was being grown in what evolved to become Peru as far back as 10,000 years ago (Schmid 2007). A thoroughly researched article in a reputable British newspaper, The Guardian, reported on the work of a team of Vanderbilt University anthropologists, who uncovered remains of peanuts and cotton, dating back to 7600 and 5500 years in Nanchoc Valley, northern Peru (Schmid 2007).

Another ancient innovation with far-reaching implications for efforts to accommodate nature in built space, is the creation of empires. Three of the region's ancient empires readily come to mind. These include, the Maya, the Aztec and the Inca. The Maya exercised dominion over a vast region, encompassing the areas currently occupied by the jungles of Mexico and Central America. The Maya are reputed for being the earliest civilization to understand the mathematical concept of zero; and the first to develop a 365-day calendar based on the sun's movement. On their part, the Aztecs controlled Mexico from the 1200s to the early-1500s. The empire's seat was an island city of mainly marshland named Tenochtitlan. This location was selected for strategic reasons; it was, by its very nature as island, shielded from real and potential attackers. However, this presented its authorities with a different set of problems. Principal among these was the problem of meeting the islanders' food needs given that marshland is not arable. To resolve this problem, the Aztecs crafted a technique known as *chinapas*, which permitted the growing of crops on what were literally floating gardens. This constitutes one of the earliest recorded use of the technique that is currently known as hydroponics. As for the Inca Empire, it occupied an area in south America on the Andes Mountains. The empire's mountainous locale posed enormous challenges regarding nature in built space. In particular, there were seemingly insurmountable difficulties with addressing the food needs of its population. To overcome these difficulties, the Inca engineers and agronomists developed engineering techniques that entailed the construction of stone terraces and irrigation systems. The terraces facilitated farming on the slopes of steep mountains while the irrigation canals made it possible for crops in otherwise dry mountainous land to receive their necessary doses of water.

6.3 ESM Components and Nature in Built Space in LAC

6.3.1 Politico-economic and Socio-cultural Context

The social context of Latin America and the Caribbean (LAC) has implications for nature in the built environment. On this score, the region's recent demographic trends are arguably the most impactful. In 2015, the region had a population of 626 million (UNEP Online). Thus, the population grew by six million between 2009 and 2015. The urban population has also witnessed tremendous growth in recent times. Between 2010 and 2015 the proportion of the region's total population based in urban areas rose from 79% to 80%. This makes the region the world's most urbanized. The most urbanized countries in the region, Brazil and Mexico contain as many as 81 of the region's 198 large cities, that is, cities with a population of 200,000 or more (Atlantic Council Online). Fifteen cities in the region grew from 'small-' to 'medium-size' between 2005 and 2015. Bogota, Colombia and Lima, Peru with populations of 9.7 million and 9.8 million respectively, recently became the region's two newest megacities. Metropolitan areas in the region continue to expand demographically and physically. At the same time, there has been a significant increase in the population densities. These trends have not been inconsequential. Rather, they have been at the root of problems such as traffic congestion, the scarcity of potable water, declining hygiene and sanitation conditions, and biodiversity degradation that are increasingly menacing human settlements in the region. In addition, there has been an exacerbation of pollution problems in especially the megacities such as Mexico City, Rio de Janeiro, and Sao Paulo.

Social problems in LAC are usually compounded by the lack of political stability. However, the situation witnessed some positive changes between the 1980s and 1990s characterized by relatively successful externally-propelled neoliberal reforms. These produced many gains, including increasingly democratic political systems, stable governments and a few economically prosperous citizens. These gains were, however, insufficient to erase the socio-economic inequalities; in addition, they were never matched by improvements in governance proficiency. Prominent in this connection is the lack of institutional capacity for formulating and implementing judicious urban plans; that is, plans that generously incorporate natural features.

The benefits accruing from economic growth in the LAC region have never been evenly distributed. While this growth resulted in improving living conditions for some, it has paradoxically boded ill for the poor. The following caption for a recent report by the Economic Commission for Latin America and the Caribbean brilliantly captures this paradox (ECLAC 2019, para. 1).

> Poverty in Latin America Remained Steady in 2017, but Extreme Poverty Increased to the Highest Level since 2008, while Inequality has Fallen Notably since 2000.

According to the report, the region's poverty level has remained the same at 30.2% (i.e., 184 million people) during the last decade. However, the proportion of the population suffering from extreme poverty—that is those lacking the ability to meet their daily food needs, rose from 9.9% to 10.2% (or 62 million people) during the same period. At the same time, the region maintains the dubious record of being the world's most socio-economically inequitable. These statistics have many negative implications for efforts to furnish built space with green infrastructure. For one thing, the poor, by definition, do not possess real estate or other taxable items. This substantially weakens the income generating ability of municipal governments, which depend largely on local real property taxes.

Heightened levels of poverty during the last decade have been accompanied by an increase in crime rates in the LAC region (IPS 2019). This has also been blamed on institutional ineptitude, particularly the state's inability to significantly reduce societal socio-economic inequalities. In her work on the retreat to gated communities by middle- and upper-income households in Trinidad, Mycoo (2006: 131) echoed this sentiment in the following words.

> The emergence of gated communities of the upper-income and middle-income classes can be directly traced to the failure of governments to close the growing divide between rich and poor and to solve the accompanying wave of crime and fear of violence. It is also a result of an abandonment of faith in governments' capacities to deliver and maintain infrastructure and civic services.

Gated communities typically come complete with tree-lined streets and meticulously manicured lawns. Thus, there is no shortage of green infrastructure in these communities. The contrast is, however, the case in the inner cities—the heart of LAC's built space.

One of these is the physical separation of socio-economic groups. In this regard, middle- and upper-income groups relocated from the inner cities into gated communities in the suburbs; while the poor became the dominant group in the inner cities. Consequently, while the suburbs grew greener, the inner cities became more congested, neglected and boasted very little green space. As a problem, green infrastructure deprivation reaches its zenith in squatter settlements and slums. This and related problems are a direct product of institutional failures.

The period marked by politico-economic gains in LAC was rather short-lived. Only a few countries in the region appear to continue on the progressive trajectory. The exact number of these countries is five, based on the Bertelsmann Stiftung's Transformation Index (BTI Online). The countries include Argentina, Chile, Ecuador, Uruguay, and Costa Rica. These countries serve as the region's luminaries with respect to political transformation; they have continued democratizing despite a general decline in democracy throughout the region.

6.3.2 Geo-ecological context

The United Nations Environmental Programme (UNEP) has presented up-to-date politico-economic and geo-ecological data on Latin America and the Caribbean (LAC) in its latest Environmental Assessment for the region (see UNEP 2016). The region encompasses 5 million square kilometers of arable land, and contains 60–70% of the world's life form. As a geographical region, it is richly endowed with natural resources. In this regard, it contains 20% of the world's known oil reserves, and 23% of the world's forests. In addition, the region receives 23% of the world's rainfall approximately 30% of the world's freshwater resources. Thus, the region alone contains as much as 70% of the renewable water resources of the American continent. This suggests that when it comes to water in human settlements in the region, the problem would unlikely be the sheer lack of resources. Rather, it is likely to be a function of managerial or administrative ineptitude.

An important but oft-ignored geo-ecological phenomenon with far-reaching implications for built space in the LAC region is climate change. The importance of climate change for the region is succinctly captured by Devisscher and colleagues (2018: para. 1) in the following words, "Latin America is one of the most urbanized, yet socially inequitable, and climate-vulnerable regions of the world." Two specific dimensions of climate change, extreme weather patterns, and rising sea levels, are particularly noteworthy here. These elements have already announced their presence in the region especially in island and coastal locales. Towns and cities in these locales face the most potentially devastating threats from rising sea levels. However, it is gravely erroneous to view the impact of climate change as limited to island and coastal communities. In fact, climate change is already having significant negative impacts on the Amazon rain forest, which contains the richest terrestrial diversity of organisms in the world. Rising temperatures occasioned by climate change have significant negative impacts on species such as fish that are temperature-sensitive. The higher temperatures thus far recorded have already caused noticeable changes in the distribution of such species. At best, as the World Wildlife Fund reports, some terrestrial species have tended to invade higher altitudes; at worst, they simply disappear from their current habitats (WWF 2019).

Another problem resulting not only from climate change but from urbanization and other changes in the region is biodiversity. Negative trends, particularly the accelerated decline in biodiversity are noticeable, with the following manifestations being the most pronounced: habitat loss, conversion and alteration, overharvesting, and rapid depletion of terrestrial and aquatic sources. Here, the effect of declining rainfall occasioned by warmer temperatures must be acknowledged. One of the many effects of the decline in rainfall is a significant reduction in bodies of surface and underground water. The decline in rainfall also results in a reduction of the amount of nutrient input to lakes, streams and rivers. This does not bode well for the survival of aquatic organisms.

6.4 Religion and Belief Systems

Religion and belief systems—for convenience sake hereafter, religion are important in the discourse on nature in built space for at least two reasons. First, religious institutions and beliefs explain the presence of certain natural features in built space. Examples of these include sacred bodies of water such as lakes and rivers, well-manicured cemeteries and sacred trees and forests. Second, religion and beliefs remain paramount among the known determinants of people's view of, and relationship with, nature. As important sources of a people's value, religion plays a critical role in shaping their orientation towards, and in fact, treatment of, the environment. Therefore, to understand how people in Latin America and the Caribbean view, and relate to, nature in built space, we must perforce gain some appreciation of their indigenous religions. Based on the means by which the region has been peopled over many centuries, the indigenous religions of the region include those practiced by descendants of pre-Columbian civilizations, and persons of African origin. Individuals of this latter category are descendants of enslaved Africans who were forcefully imported to the region in the fifteenth century. Their indigenous religions are among the few aspects of their culture that they preserved and jealously safeguarded. Once out of Africa, the religions experienced substantial syncretism, metamorphosed into a variety of belief systems and assumed different names. In Cuba, it assumed the name, *Regla de Ocha* or *Santeria*, in Haiti, it is known as *Voodoo,* and in Brazil, it is called *Candomblé* (Olupona 2015). These religions have been considerably adulterated in their host locales. For instance, Santeria, which means worship of the saints in Spanish, contains noticeable elements of Catholicism. Yet, their ostensible African identity remains irrefutable. According to Eurocentric accounts, these religions, like those of their African forebears, are polytheistic. In other words, they honor multiple gods or deities. This characterization was, and remains, grossly erroneous. What Eurocentric analysts misrepresented, deliberated or otherwise, as multiple gods were actually a Supreme God and his adjutants. In Christian ethos, these would be the Almighty God and his angels. Thus, indigenous religions cannot be considered any more polytheistic than Christianity, which is well known for mandating the worship of God Almighty and reverence of angels and saints.

The worship of the Supreme God and his adjutants is one of the few attributes that indigenous religions with roots in Africa share with Christianity. For instance, the adherents of Candomble and the Yoruba religion it originated from believe in an Almighty God called Oludamare, who is served by lesser deities known as Orishas, Voduns and Inkices. Orishas are believed to be the adherents' ancestors who were long gone but have undergone the deification process.

Nature and especially the natural environment are of centrality to adherents of these religions. The spirits of Orishas are believed to be anchored in natural elements, particularly the earth, trees, leaves, water and wind. Also, the adherents believe in the healing power of nature. Animals command a lot of respect in these religions. For instance, Candomble doctrine holds that animals are capable of transmitting life-saving messages from Orishas to humans. This is because, according to the doctrine, humans lack the ability to understand non-verbal communication.

The belief systems and cultural practices of the descendants of pre-Columbian civilizations of the America's (North, Central, and South) and the Caribbean remain in vogue today. The indigenous people of this vast region are variously labelled Native Indians or Aboriginals in Canada. The appellation, India is a misnomer as it was coined by Christopher Columbus, who, upon encountering the first members of the region's native population erroneously thought he had arrived India in Asia. The Americas is a vast area and has never had a single indigenous religion. Rather, there have always been multiple indigenous religions throughout the region. However, they share many features. These include a number of stories and legends, the most prominent being the creation story, myths that are mythological. In addition, they have Shamans comprising traditional healers, ritualists, singers, mystics, lore keepers, medicine men and women (New World Encyclopedia 2019). The spiritual practices and doctrines are described by words in indigenous languages, and never taught to outsiders. The indigenous religions of the region may be better characterized as spiritual in contrast to religion in the conventional sense. These practices were gradually supplanted by Christianity especially Catholicism in the case of Central and South America, and Protestantism in present-day United States and Canada subsequent to the European conquest.

The indigenous people of the region can be divided into two major groups. These include the mountain people and the people of the forest areas. The forest people have practiced organized agriculture for many centuries before the European conquest. The forest people have always existed as hunters and small-scale farmers. Throughout time, both groups have depended on natural herbs for preventative and curative therapy. Accordingly, 'shamanism' and 'herbalism' continue to be vital in the region. These indigenous people practiced collective as opposed to individual land ownership. This latter was introduced by Europeans as part of the colonial enterprise. This difference in land ownership model remains a source of anti-government demonstrations and agitation in Latin America. Protests against governments in the region to recognize and legitimize collective land ownership as a land tenure option are commonplace. Apart from seeking recognition of collectively-owned tribal lands, indigenous people in the area have opposed environmentally damaging activities on their lands by outsiders whose interests center mainly around profit. An aspect of the cultural practices of indigenous peoples of Latin America with implications for the natural environment has to do with agriculture. In this case, they practiced 'slash-and-burn', which is often the object of fierce criticism. Critics contend that it is harsh to the environment. However, such criticism seems to be a function of ignorance of the genre of 'slash-and-burn' practiced by indigenous peoples. Typically, these individuals clear a small portion of forest land at a time. This area is then farmed for a short period—about a year or two. Thereafter, the farmers typically move to another piece of land. This allows the farmed area to fallow and heal itself without any environmental damage.

Another notable element of indigenous religion with positive implications for environmental stewardship is the proclivity for maintaining sacred sites. Sacred sites are of two variants, namely man-made and natural. The man-made typically belong to organized religious bodies. Examples of these include the following (Grim n.d.):

- Temples of Heaven and Earth in Beijing;
- The wailing Wall and Dome of Rock of Jerusalem;
- Saint Peters Basilica in Rome;
- Muslim Pilgrimage sites;
- Monuments of the Buddhist remains at Borobundur;
- The Hindu—Buddhist sites at Angkor in Cambodia;
- Sacred mountains in China.

Sacred sites of the natural variant are of centrality for the purpose of the discussion in this book. These include bodies of water, areas of land and places of biodiversity concentration that possess special spiritual significance to indigenous people and others with whom they share religious or cultural affinity. Before delving deeper into the subject of sacred natural sites in indigenous societies, it is necessary to underscore the fact that in such societies, the manifestation of space—time has as its primary focus is not the human but the cosmological forces experienced in the land (Grim, n.d.). The sacred places are typically natural objects such as rivers, islands, prairies, hills, valleys and mountains; trees, grass, animals and aquatic life. For centuries, indigenous religions have acknowledged and asserted their ancestral relationship to the Earth through these entities. In the United States, colonial authorities displayed a wanton disregard of indigenous rights to sacred sites as they moved speedily to bring every centimeter of land throughout the country under government control. As part of this conquest and domination enterprise, were efforts to coercively convert members of America's indigenous population to Christianity. This resulted in the relocation of members of this population or Native Americans—in some cases, forcefully—from their homelands and sacred sites. However, despite being relocated to other locales, the sites remained and continue to be sacred and treated with reverence. A place can become sacred for one of several reasons. For instance, as noted by Native American Netroots, such a place may be conferred the 'sacred' status because it is the locale where a religiously notable act occurred (Netroots 2019). The act may involve a feat of creation that occurred before human existence; or the locale in which ancient ancestors are believed to have executed a task of significance in the history of the indigenous people. Consider the following tale regarding the conferment of the 'sacred' status on Bull Lake in Ontario, Canada. The tale as recounted in the Native American Netroots, tells of a young hunter who, once upon a time, relentlessly pursued a white buffalo until it leaped into the lake (Netroots 2019). Indigenous members of the Shoshone tribe believe that the spirit of the white buffalo continues to live at the bottom of Bull Lake to date. Consequently, Bull Lake was conferred the 'sacred' status by the Shoshones' forebears. Here, I hasten to note that once any place has been conferred the 'sacred' status, it remains sacred ad infinitum. Hence, the fact that Bull Lake remains a sacred place to date. It remains a place to visit for members of the Shoshone tribe desirous of making a spiritual connection with the spirit world. Typically, this ritualistic visit requires the devotee to spend a night on the lake's banks.

Recently, there have been conscious efforts to recognize and re-instate indigenous rights over sacred lands and places in the United States. These efforts culminated in the enactment of the American Indian Religious Freedom Act four decades ago. This Act recognizes the freedom of Native Americans to practice their indigenous faith as well as the need to preserve their sacred sites. There are many such sites, which some have likened to (Christian) churches, in the United States (Conversation 2019). As an equivalence to churches, sacred sites constitute places of worship, healing and re-connection with one's forebears. Examples of sacred sites in the United States include, Glen Cove (Ssogoréate) Sacred Burial Sites in California; Bears Ears National Monument in Monticello, Utah; Bear Butte in South Dakota; the sacred hills in Nebraska; the sacred hills of North Dakota; the Sweetgrass Hills in Montana; and the San Francisco Peaks.

6.5 Guaranteeing Food Security Through Urban Agriculture

6.5.1 Importance of Urban Agriculture in LAC

As noted above, Latin America and the Caribbean is the most urbanized region in the world. This magnifies the importance of urban agriculture for sustainable development in the region. The Food and Agriculture Organization (FAO) of the United Nations is cognizant of this, and has recently undertaken a number of insightful studies of urban and peri-urban agriculture (UPA) in the region over the years. Many informative pieces of information can be gleaned from these studies (see e.g., FAO 2019). A recurring theme in the studies is that urban and peri-urban agriculture (UPA) involves many participants and plays a critical socio-economic role in Latin America and the Caribbean (LAC). Urban and peri-urban agriculture (UPA), as the name implies, includes two main prongs. The one relates to farming within urban areas, while the other has to do with farming activities at the outskirts or peripheries of urban centres. The main functions of UPA have been identified to include the following: supplement the food supply of households, create employment opportunities, and contribute to environmental resilience. However, it is difficult to appreciate the real or potential contribution of UPA in any country without some understanding of its magnitude or extent.

Some evidence of the extent of UPA is provided by the fact that participants are drawn from disparate demographic and geographic milieus, including major urban agglomerations such as Mexico City, mid-size cities such as San José del Golfo in Guatemala, prosperous capitals such as Belo Horizonté in Brazil, as well as overcrowded communities of internally displaced populations such as those at the outskirts of Port-au-Prince in Haiti. The intensity of participation is bolstered by statistical evidence. For instance, 40% of households in Cuba, 20% in Guatemala and Saint Lucia respectively, are involved in one or another UPA activity. Numerically, as many as 8500 households in Bogota, Colombia, 25,500 in Port-au-Prince, Haiti, and 1.4 million households in Nicaragua and Guatemala, are involved in UPA activities. Havana, Cuba, whose UPA program is described in greater detail in Box 6.1, ranks at the top of the region's 'greenest cities.' As many as 90,000 residents of this city practice some form of UPA. All accounted for, 54% of Cuba's population participate in UPA. In comparison, 26% of the population in Antigua and Barbuda, 54% of Bolivia's population and 68% of the population of the Dominican Republic are involved in urban and peri-urban agriculture. Quito, Ecuador, which ranks as the region's second greenest city, boasts several UPA activities, including 140 community gardens, 800 family gardens, and 128 school gardens. The benefits of these activities to households are not limited to the quantitative and qualitative dietary supplements they derive from the resultant food. Rather, they also include savings of funds that would have gone towards purchasing food for the household. By some estimates, such savings amount to as much as one-fifth of household income in the region (FAO 2019).

Urban farmers in the region cultivate a variety of food crops of their choice; although this choice is conditioned by ecological factors and spatial limitations.

Thus, they are limited to only those food crops that can thrive in the region's temperate climate and in tight spaces such as people's backyard, porches, front lawns, open boxes or tins, and so on. The common urban farming units are individuals, households, extended families and communities. Family gardens happen to be commonplace in some countries in the region, particularly mainland countries like Colombia, Ecuador and Peru; and in some Caribbean countries such as Cuba and Jamaica. Urban farm products are also widely variable by sub-region and country or city; typically, they include eggplants and okra (e.g. Antigua and Barbuda), carrots and coriander (e.g., Tegucigalpa), broccoli and quinoa (e.g. Quito in Ecuador), spinach and strawberries (e.g., Bolivia's *altiplano*). To be sure, not all urban farmers limit themselves to the production of food crops; some are involved in animal husbandry. Sample cities known for farming activities of this latter genre include El Alto, where some families raise guinea pigs; and the outskirts of Mexico City, where the rearing of goats, sheep, rabbits and birds is ubiquitous. Mexico City's situation is especially worthy of note because it is facing an acute land scarcity problem. Here, informal settlements are increasingly expanding into land previously used for urban farming. This has led to, among other things, a water shortage problem occasioned by the over-exploitation of aquifers. The land scarcity problem notwithstanding, Mexico City's urban farmers use as much as 22,800 ha of land to produce 15,000 tonnes of vegetables. On the outskirts of Lima, Peru's capital city, urban farmers grow short cycle vegetables on 5000 ha of irrigated land. Urban farming is also thriving in Jamaica. In its most vibrant urban centre, Kingston City, fish rearing is commonplace; it involves young people who breed tropical fish for exportation to North America.

Although those involved in UPA originate in all walks of life, most are drawn from low-income households. Worth noting is the fact that the low-income individuals in UPA often count on it as their sole source of income. In contrast, those of middle- and upper-income households practice UPA as a hobby or consider it a means of subsidizing their nutrition intake. Also, women who are single household heads dominate urban and peri-urban farmers. As the FAO report noted, they constitute 90% of urban and peri-urban farmers in Managua, Nicaragua; 86% in Port-au-Prince, Haiti; 70% in Belize City, Belize; and 25% in Quito, Ecuador. The report also identified additional benefits of UPA to include improved access to food, and diverse diets (FAO 2014).

Urban and peri-urban agriculture (UPA) is not without challenges in LAC. Paramount among these are the following: lack of space, poor soil quality, unreliable water supply, high cost of inputs, and lack of quality seeds. These problems have caused a number of problems beyond the UPA domain. For instance, in Lima, Peru, water scarcity has caused UPA farmers to resort to irrigating their farms with polluted waste water—an act that can threaten human health. Other common challenges include the lack of credit for the acquisition of tools and processing equipment/technology to add value to farm products, and limited access to markets.

An important but oft-ignored requirement for the success of UPA is government support. As Box 6.1 shows, the support of the Cuban state is an important, if not the most important, reason for the illustrious success of Havana's UPA program. Cuba

Box 6.1: Urban and Peri-Urban Agriculture in Havana, Cuba

Cuba has a history of food self-sufficiency that dates back to the immediate-post-revolution era that began in 1959. Two decades later, in 1980, the country had a population of 10 million but produced enough food to feed 40 million people. However, this situation was drastically altered as of the 1990s subsequent to the demise of the Soviet Union. The Soviet break-up resulted in a loss of Cuba's main loyal trading partner and its main source of fossil fuel, mineral fertilizers and pesticides. One outcome of this was a reversal of its fortunes in the agriculture sector. The problem was aggravated by the U.S. trade embargo. For the first time, Cubans faced dire food shortage that necessitated food rationing.

To arrest this situation, authorities in the country's capital and largest city, Havana, encouraged residents to take up urban and peri-urban farming (UPA). This was essentially a call for Havana residents to supplement their food supply and save on food expenditure by planting crops in any available space within the city. The call yielded exceedingly positive results as people throughout Havana took up farming by raising crops on vacant lots, in backyards, and soil-filled containers mounted on porches and rooftops. Initially, the food yield was nothing to write home about. However, over a short time, and after the government threw its full weight behind the UPA initiative, the yields increased by leaps and bounds. By 2012, according to the Food and Agriculture Organization (FAO 2014), families were already cultivating as many as 89,000 backyards and 5100 plots of less than 800 square metres. Families mainly grow fruits, vegetables and condiments. In addition, they raise small animals, including poultry and guinea pigs for household consumption. If nothing else, Havana's success underscores the importance of government support for UPA initiatives. As the Food and Agriculture Organization (2014: 10) observed, through its support, the Cuban government had transformed what was a spontaneous response 'to food insecurity, to a national priority.' In doing so, Cuba, through its premier city, Havana, became 'a pioneer in a worldwide transition to sustainable agriculture that produces more with less.'

The implementation of UPA policy in Havana entails two main prongs. Both fall under the auspices of the national government. The first prong concerns urban agriculture while the second relates to the peri-urban areas. The strong support of the Cuban government for urban and peri-urban agriculture is evident. Witness for instance, the fact that it established several provincial offices and technical departments throughout the country to support this form of farming. In Havana alone, the government established a provincial service the Havana Provincial Office of Agriculture, seven provincial technical departments and 15 municipal offices. These offices are charged with the

(continued)

Box 6.1 (continued)

responsibility of providing technical and logistical assistance to urban farmers in Havana. In addition, the Cuban government has made provision to grant vacant land free of charge to women and youth interested in undertaking urban agriculture project. Furthermore, the government has created a Technical Advisory Board that represent agricultural research institutes. Other important and noteworthy features of the Havana's urban and peri-urban agriculture sector include five agricultural enterprises. These manage some 700 crop farms, 170 cattle farms and 127 tree production units, two provincial companies specializing in pig and livestock production, 29 agricultural cooperatives, and 91 credit and service cooperatives that grow flowers and vegetables and raise small animals.

Another way of appreciating the magnitude and extent of Havana's urban agriculture is to understand the total area assigned to the UPA sector. This area is estimated to be about 35,000 ha, that is, half of the area of the entire City of Havana. Also, the extent of the UPA sector can be understood by appreciating its sheer production capacity. In 2012, according to the FAO (2014), this capacity was 63,000 tonnes (metric tons) for vegetables, 20,000 tonnes of fruits, 10,000 tonnes of roots and tubers, 10.5 million litres of cow, buffalo and goat milk, and 1700 tonnes of meat. Yet another meaningful measure of the extent and size of the UPA sector in Havana is by taking stock of the number of households involved in the sector. In 2014, 90,000 households were involved in urban and peri-urban agriculture in the city.

Source: FAO (2014)

is not alone as other national and municipal governments in the LAC region have been active in efforts to support UPA. These efforts have assumed many forms. For instance, some governments have helped urban and peri-urban farmers to market their surplus products. A common approach to this has involved the establishment of food markets in strategic locations—that is, locations within easy access for urban farmers and their real and potential clients. There are 14 such markets in the Ecuadorian capital, Quito. Each of the markets holds once per week. They are typically located in low-income districts. In 2012, these markets brought in US $76,000 to Quito's economy. However, no discussion of farming in Quito can be deemed complete without some mention of the city's role as a regional player in UPA. The city served as the venue of the April 2000 meeting of nine LAC countries that crafted the Quito Declaration. This is the first of its kind; it called for the adoption of UPA as a strategy to combat poverty, food insecurity and environmental degradation. Apart from serving as the site where the Declaration was signed, Quito continues to be a leader in its commitment to support urban farmers. The government provided the city's 32 urban parishes seeds and seedlings, equipment, management and training, and poultry, guinea pigs, and bees as well as other valuable input.

6.5.2 Extent of Institutional Support

Two important state-induced shifts that have moved urban planning, albeit slightly, away from orthodoxy, have occurred in LAC since the nineteenth century. The first entailed the adoption of the garden city model of Ebenezer Howard (see Box 6.2). The second is of more recent vintage, and occurred between 2007 and 2008. It involved the modification of urban planning laws and regulations to permit agricultural activities in urban areas. With the UN Food and Agriculture Organization (FAO) at the helm, this modification permitted the use, for farming purposes, of vacant space in urban and peri-urban locales. Such vacant space that have typically gone unused include vacant lots, backyards, building roofs, and spaces along high tension electric cable right of way. Knowledge of the fact that such spaces have always existed but hardly used for farming because of stringent urban planning laws, dictated a need for flexibility on the part of these governments. In addition, national and municipal governments recognized the need to actively support urban farming as a means of contributing towards efforts to promote food self-sufficiency and security. This explains the heightened level of active support that UPA has

Box 6.2: The Garden City Idea in LAC

Few planning ideas imported from Europe are as important as the garden city model in Latin America and the Caribbean. The first proposal to employ the model was by French urban planner, Alfred Agaché. He led a French planning team that had been invited to craft a new plan for the City of Rio de Janeiro in the late-1920s. As a devout member of the English Garden City Movement Agaché considered this invitation as an opportunity to showcase the garden city plan and promote its ideals. He fervently believed that the model constituted the best answer to local authorities' request for a plan that could afford the city of Rio the spatial order it badly needed and a degree of aesthetic appeal befitting a great modern city. However, the original impetus for the garden city model as initially proposed by Ebenezer Howard was more to promote green space—farmland, parks, and natural vegetation—and little to do with aesthetic appeal. Accordingly, the types of plans that Agaché and others of his ilk promoted in LAC would have likely been rejected by Howard as failing to reflect his ideals. For instance, Agaché's plan for Rio de Janeiro proposed a distinctly segregated spatial structure that spatially separated people by socio-economic class. In so doing, the plan promoted an attribute of modernist urban planning that the garden city model was intended to eliminate, namely the excessive dependence on motorized modes of transportation. More importantly, the Agaché plan failed to emphasize as much nature—gardens and ancillary infrastructure—in urban areas as Howard would have preferred.

Source: Njoh (2015)

received from national and municipal governments in LAC during the last decade. These governments are ideally placed to resolve any conflict among land uses because of their dominant role in conceiving, formulating and implementing land policies.

More importantly, the state is the principal land owner in most countries, particularly those in Latin America and the Caribbean (LAC). In most cases, governments serve as managers not only of the land but also of solid wastes and water. In this capacity, the government can deny or grant permission to use resources necessary for urban farming. A few countries in LAC are already taking steps towards creating a conducive environment for the growth of UPA. In its 2014 study of urban and peri-urban agricultural activities in the region, the Food and Agriculture Organization (FAO) found that a dozen countries in the region have already enacted policies specifically designed to promote urban farming. Eight of these are in the Caribbean, and some, such as Cuba, have had these policies in place since the 1990s. As shown in Box 6.1, Cuba's program is particularly noteworthy because of its prioritization by the government. It was initiated by establishing a network of agricultural supply stores, municipal seed farms, composting units, veterinary clinics for breeding biological pest control agents. In addition, the Cuban government provides loans to enable farmers procure insurance and facilitate production.

Another country that prioritizes support for urban agriculture is Brazil. This country runs a program known as the National Zero Hunger Policy. This program constitutes the channel through which the Brazilian government supports the country's UPA. The support contains elements such as direct government technical and financial input into efforts to build necessary infrastructure. Past beneficiary activities in the country include, the acquisition of land for, and construction of, farmers' markets; defraying the cost of training for gardeners; purchase of vacant urban spaces for farming; and covering the cost of foregone revenue for reduced taxes on urban land set aside for farming.

Bolivia also counts among countries whose government has afforded UPA heightened attention. The government of this country is on record for adopting a national food and nutrition policy that endorses UPA. In addition, the country's Ministry of Productive Development and Plural Economy, with the assistance of the Food and Agriculture Organization launched a national program to encourage UPA in 2014. Apart from its proactive UPA-promoting policies, Bolivia is of regional importance with respect to urban farming. One of its main cities, Cochabamba, served as the venue for the ratification of the Organization of American State's "Declaration of Cochabamba" of June 2012.

In Antigua and Barbuda the government has policies in place to support UPA and especially backyard gardening. The government's strategies in this regard have involved the use of extension workers and community facilitators. Also, authorities have made other inputs such as the provisioning of seedlings, and fruit trees free of charge or at a significantly devaluated cost. In Guatemala, the state, through its Ministry of Agriculture, Livestock and Food has established a Department of Agriculture, which it has charged with the responsibility of training and arming farmers with tools and other necessary inputs. In Argentina, the country's National

Institute of Agriculture and Livestock Technology has been very active in UPA. One of its most notable activities in this connection has been the creation of gardens in major urban centres throughout the country.

This has taken place under the auspices of the ministry's Pro-Huerta Gardening Program. The program is credited with creating 8000 community gardens, 7000 school gardens and half a million family gardens. Another testament to the Argentinian government's commitment to UPA resides in Rosario. Here, as reported by the FAO (2014), the municipal government assigned as many as 25 agronomists and allocated the sum of US$380,000 annually to urban agriculture. The national government, through Pro-Huerta, complemented the efforts of Rosario's municipal government by providing local urban farmers necessary training, seedlings, and tools.

Especially worthy of note as far as efforts to promote UPA in LAC go is the fact that many countries, precisely 17 out of the 26 studied by the FAO (2014), have at least one government ministry in charge of urban farming activities. The responsibilities of these ministries range from overseeing and regulating to providing material and logistical support to urban farming activities. Even in countries such as Colombia, Peru, and Ecuador that have no formal UPA policies, urban agriculture has been mainstreamed. Thus, it is safe to conclude that UPA is alive, well and strong in major cities throughout the LAC region.

6.6 Promoting Nature in Built Space Through Urban Forests

There has been a shift in the zoning paradigm that has dominated urban planning since the eighteenth century. The shift has been occasioned by new thinking that rejects the highly compartmentalized land use spatial configuration of orthodox or modernist urban planning. Instead, cutting edge thinking in planning dictates a flexible spatial configuration that permits mixed land uses and treats nature as a bona fide feature of the urban landscape. For Latin America and the Caribbean (LAC) region, this invariably requires extensive retrofitting especially in the case of the old cities whose nucleus was established by colonial authorities in the sixteenth century. Post-colonial authorities expanded these and created new ones of their own during the heydays of the region's economic boom of the 1900s. Although the urban development initiatives of this latter period took place during the post-colonial era, they were influenced just as much by European ideas of urban order and structure as their predecessors of the colonial epoch. As I observed elsewhere, the 1900s are particularly worth noting in the urban history of LAC because of the rapid growth experienced by cities in the region (Njoh 2015). The growth had several dimensions prominent among which was the sprawling of urban residential areas into suburbs. This compelled members of the middle-class who previously occupied the suburbs to seek new locations further removed from the inner city. One factor that exacerbated this problem is the advent of the automobile in the region. In an attempt to address this problem, authorities invited urban planners who were either Europeans or local experts educated in Europe. When it came to responding to the need for open and/green space, the planners were unified in recommending the garden city model, which was crafted in 1898 by a British stenographer-cum-urban planner. The model is best-known for its emphasis on the need to entertain green space, particularly space for agriculture and cognate activities as an integral part of the built environment. At the forefront of the movement to promote the model in Latin America was a Frenchman by the name of Alfred Agaché (Njoh 2015). Agaché and other planners that adhered to European planning principles in LAC at the time were interested in developing self-contained urban communities that are circumscribed by greenbelts. Thus, the model is of interest to us here because of its 'green content.' On the ground in LAC, the model was slightly modified and attributed a different moniker, namely the 'garden suburb.' This was essentially to accentuate the fact that unlike its use in its locale of origin, the model was employed mainly to promote spatial order in suburbs in LAC. Beneficiaries of the model in the region included the first *colonias* of Porfirio Díaz's Mexico City, the 1890s area of Higienópolis in São Paulo, the *urbanización* El Paraíso in 1900s Caracas, and Havana's *Vedado*. A relatively later use of the model was in the preparation of the *Jardin America* in Sao Paulo in 1915. Since then, hyper-urbanization and negative trends in the economies in the region have resulted in the devaluation of initiatives to develop and maintain nature in built space in the region.

Recently, there have been renewed interest in these initiatives. The initiatives particularly seek to maintain as mush of the region's rich biodiversity in built space as possible. In this regard, national, regional, and international authorities are taking serious steps to combat climate change and deforestation (GRC 2017). For instance, authorities in Colombia have launched a sustainable development fund of US$210 million, that targets urban greening *inter alia.* In addition, many countries in the region have accentuated efforts to preserve underwater treasures. Regional leaders in these initiatives include Mexico, which has created the Revillagigedo National Park. Enclosing a total of about 15 million hectares (approximately 37 million acres), this is the largest marine park in North America. On their part, Costa Rica and Ecuador recently signed an accord, the Coral Reef Life Declaration. The timely and all-too-important accord has the avowed aim of preserving healthy coral reef ecosystems. Furthermore, many municipal authorities in the region, especially those in Argentina and Chile have banned the use of plastic bags, and their disposal in any body of water. More importantly, many countries in the region have registered some success in efforts to conserve terrestrial landscapes and freshwater resources. These efforts have benefited from the goodwill of generous non-governmental organizations (NGOs) and other entities at the national and international levels. In this regard, the government of Chile benefited from a donation of more than 400,000 ha (about one million acres) of land in Patagonia. By some reckoning, this gift would permit the Chilean government to create a network of national parks as large as Switzerland (GRC 2017).

Despite their importance, none of the aforementioned initiatives has as much of an impact on efforts to create and preserve nature in built space as green infrastructure development. The role of green infrastructure in improving the aesthetic appeal of built space and ameliorating the quality of life for their residents has been widely acknowledged. However, Latin American cities score considerably below the average for European cities with respect to access to green space (Terraza Online). Northern European cities are, on average, furnished with 540 square feet of green space per resident. In comparison, central American cities have less than 121 square feet per person. More striking is the fact that green space is inequitably distributed in the region. This often results in residents of wealthier neighborhoods having parks and/or other recreational space within walking distance. In contrast, residents of low-income neighborhoods are never within close proximity to such facilities. Consequently, to access green space, the poor in LAC require motorable transportation, which entails extra cost. Thus, when it comes to access to green space, the LAC region constitutes a classic case of spatial injustice.

There is growing evidence suggesting meaningful efforts to increase the stock of green infrastructure in LAC. In this regard, Costa Rica ranks among the region's leaders. However, the country reportedly experienced some setbacks in the 1990s (UN-Habitat 2009). This is when the country lost some of its forest land to development. Nevertheless, Costa Rica's activities in the ecology domain are laudable in both qualitative and quantitative terms. For example, the Costa Rican government planted five million in 2007 and pledged to plant seven million more in 2008. This translated to 1.5 trees for each Costa Rican at the time. In addition, the government committed to stop deforestation. Another LAC country that has undertaken noteworthy initiatives in green infrastructure development is Chile. Vasquez et al. (2016) have conducted a study of such initiatives in that country's premier city, Santiago. This study and the results it recorded are summarized in Box 6.3.

Box 6.3: Green Infrastructure Development in Santiago

Santiago has eight million inhabitants, which makes it the most populous city in Chile, and the seventh most populous in the LAC region. The city has grown by leaps and bounds during the last three decades. This rapid growth has placed the city on a collision course with nature, particularly the natural vegetation. Especially noteworthy in this regard is the fact that the man-made changes and the damage to the natural environment this engendered caused major interruptions of wind/ventilation in the corridors of sediment and water flows. Above all, this exposed the population to environmental hazards. This dictated the need to develop and maintain "a green infrastructure interconnected network of proper functional ecosystems based on key landscape elements—such as wetlands, forests, creeks, lakes, rivers, and others" (Vasquez et al. 2016: 1411). The task of developing a green infrastructure system for Santiago proved especially challenging for two main reasons. First the city system is highly fragmented. Second, the Chilean government has "no institutions capable of addressing the complexity of planning and development of green infrastructure system" (Vasquez et al. 2016: 1414).

Source: Vasquez et al. (2016)

Municipal authorities in Medellin, Colombia have also been preoccupied with efforts to 'green' their city (Bergvall and Dahl 2015). In this regard, the authorities have taken actions to employ green belts as a means of separating land uses. At the local level, urban planners and policy makers have wrestled with questions regarding the relation between urban and rural settings, and the inevitable interaction that occurs between them. As discussed earlier, the use of green belts as an instrument to control growth and contain sprawl is not novel. In Medellin, the instrument was employed not only for this avowed purpose but also to mediate the relationship between two adjacent communities—one formal and the other informal. Together, these communities constitute the metropolitan area of Aburra Valley, which is divided into two segments by the Medellin River. At the heart of the Metro-Area are four densely populated municipalities, namely Medellin, Bello, Itagul and Emigado, which form the urban zone. Medellin's growth is restricted by nature, particularly topography. The city lies in a valley, which is encircled by steep hills. It is confined to the accessible land along the river. The city's planning department has been busy steering the new growth to the flat land at the city centre. Fragile eco-systems characterize the hillsides, which are inhabited mostly by low-income households. The settlements on the hillsides are of the informal variant.

Municipal authorities have, for quite some time, been pre-occupied with efforts to control the spread or expansion of the informal settlements. Earlier growth control strategies have included upgrading the slum settlements—a strategy that proved ineffective. The green belt was the latest that the municipal authorities elected to employ in an attempt to control urban expansion and improve the environment. The project comprised some 75-km-long nature area circumscribing the city. The aim of the green belt or *cinturon verde* is to separate the urban from the rural areas. Thus, it served as the urban—rural border. The avowed purpose of this is to create conditions and opportunities for integrated and sustainable development in the rural-to-urban fringe.

Natural Heritage in LAC Places classified as 'natural world heritage sites' assume heightened importance in discussions of built space in Latin America and the Caribbean (LAC). This is especially because of the region's exceedingly rich history and level of urbanization, which is the highest in the world. This means, *inter alia* that, as part of the stock of natural resources in the built environment, these sites are under constant threat from urbanization and human activities writ large. In fact, analysts are unified in concluding that tropical biomes such as those characterizing LAC are facing enormous and ferocious threats from human activities (Eyre 1990). Thus, discussing these sites in the context of built space is apropos. Tilman Jaeger (2013) has done a fine job analyzing these sites, which constitute critical attributes of the landscape of the LAC region—widely recognized as the world's most biologically and culturally diverse. Out of the 17 'megadiverse' countries as classified by the United Nations Educational, Scientific and Cultural Organisation (UNESCO), as many as six are located in the region. The region contains 39 UNESCO natural World Heritage properties classified according to a number of natural World Heritage criteria. They cover almost 37 million hectares of land and sea in the region. The 39 properties are distributed as follows: 22 in South America; 13 in Central America and Mexico; and 4 in the insular Caribbean. Fourteen of the properties contain marine reserves as well as terrestrial areas. In addition, four terrestrial properties include coastal areas although they are typically classified as forest reserves. Prominent among these are two reserves, the Atlantic Forest South-East Reserves, and Discovery Coast Atlantic Forest Reserves. Others include the Rio Platano Biosphere Reserves in Honduras, the Darien National Park in Panama. A good many of the reserves include marine or at least coastal areas, and are regional in extent. Examples include the Eastern Pacific Marine Corridor, the Galapagos Islands in Ecuador, Malpelo Fauna and Flora Sanctuary in Colombia, Coiba National Park in Panama, Cocos Island National Park in Costa Rica, and Gorgona national Park in Colombia. One of the properties, the Talamanca Range-la Amistad Reserves/ La Amistad National Park straddles the boundary between, and is shared by, Panama and Costa Rica. Other natural heritage areas of the transboundary variant in the region include the Tikal Forest in Guatemala, which is part of the larger tropical forest that stretches across parts of Belize and Mexico. In addition, there is the Pantanal Conservation Area in Brazil, which is part of South America's wetlands that extend into parts of Paraguay and Bolivia. Some of the natural heritage properties have been identified as endangered and in need of special attention. Included on this list are, the Rio Platano Biosphere Reserve in Honduras, Los Katios National Park in Colombia and the Belize Barrier Reef Reserve System in Belize.

6.7 Energy and the Built Environment in LAC

For a long time in the history of the Latin America and Caribbean region, the supply and distribution of energy has been the responsibility of the state. In the case of electricity, for instance, state and parastatal agencies were the main suppliers. Thus, in a typical LAC country, there was a national interconnected grid linking "isolated power systems served in many cases by private utilities, and assigned to new state-owned enterprises" (Dussan 1996: 1). These state-owned enterprises (SOEs) were essentially monopolies in the energy domain. They were responsible not only for the supply, but also the transmission and distribution, systems. While this model was relatively effective before the region's population and built space began experiencing rapid growth and expansion, it has proved increasingly inadequate since then. The model reached its breaking point in the early 1980s. This is when the demand for utility services, especially electricity, occasioned by high rates of population growth and economic prosperity, began its upward trend. Thus, the period was characterized by high rates of service penetration, large economies of scale in generation projects, and strong economic development. In addition, the energy sector was prioritized as a beneficiary of financial and technical support from the government.

Over time, this model proved unsustainable especially as government resources were drying up and the public debt was astronomically rising. This prompted a need for badly needed reforms. Chile was at the forefront of the march towards energy sector reform in the region. In this regard, Chilean authorities began taking active steps in the mid-1980s to reform the nation's power sector. Initial actions on this front included the decoupling of energy generation, supply and distribution systems, the creation of a competitive environment, and above all, the privatization of SOEs. Following in the footsteps of Chile, Argentina, Peru, and Colombia embarked on efforts to reform their energy and utilities sector in the early-1990s. Other countries in the region soon embarked on similar initiatives. While these reforms were substantive, they proved ineffective in addressing the region's growing and increasingly complex energy problem. Paramount among the shortcomings of the reform initiatives is the fact that they paid little attention to environmental implications of energy generation, supply and consumption. Consequently, authorities in the region are once more pre-occupied with efforts to reform the energy sector in general and non-industrial energy supply and distribution in particular.

The LAC region contains oil exporting countries and many that rank high among energy producers and consumers by global standards. Venezuela, for example, counts among the world's oil producing leaders. It is a member of the Organization of Petroleum Exporting Countries (OPEC). The region also ranks high as an energy consumer. As consumers, a good number of countries in the region continue to depend on biomass energy sources. For instance, Guatemala, El Salvador, Honduras, and Nicaragua, fuelwood remains the dominant source of energy. While this may be perceived as positive news for the environment because it implies less demand for fossil fuels, it bodes ill for forests and forest resources. Hydrocarbons, which account for 80–90% of its total energy supply remain the dominant source of energy in the region. Some countries in the region, including Argentina, Mexico, Venezuela and Ecuador rely on very little to no biomass use. This suggests that the region faces problems of sustainability with respect to energy supply and availability. By some estimates, the region would have to almost double its currently installed power capacity to about 600 gigawatts (GW) to meet its projected population by 2030 (Vergara et al. 2014: 1).

At any rate, the region is at the forefront of efforts to promote the use of renewable energy. It is generally agreed that establishing quantifiable global objectives in the renewable energy sector always proves to be a challenging undertaking. For example, during the World Summit on Sustainable Development that was held in Johannesburg, South Africa in 2002, participants conceded that establishing global targets and deadlines for renewable energy was a difficult if not impossible proposition. The degree of difficulty related to this task notwithstanding, the LAC region managed to establish concrete renewable energy targets. Continuing from where they left off in Johannesburg, the LAC renewable energy authorities met in Brasilia in October 2003 for the Brasilia Regional Conference on Renewable Energy. The preeminent item on the agenda of this meeting was to establish in concrete terms, specific targets that were established in Johannesburg. The Brasilia Conference was also on record as the first to ever bring together authorities and representatives of LAC ministries of the Environment and Energy. A major accomplishment of this conference was the approval of the Brasilia Platform on Renewable Energies. The platform established the following as its main points:

- To promote initiatives aimed at achieving the goals established at the Latin America and Caribbean Initiative for Sustainable Development; and
- To ensure a speedy transition to renewable energy in the region.

No discussion of issues driving energy supply and demand in the LAC can be deemed complete without mention of the rapid changes occurring within the region. The most remarkable of these are occurring in urban areas. These areas are experiencing rapid growth in both demographic and physical terms, which invariably calls for a corresponding increase in energy supply. Some analysts contend that improving energy efficiency alone is insufficient to address the region's energy need now and in the future (Vergara et al. 2014).

The region has made many noteworthy achievements in the energy sector in the recent past. According to the International Renewable Energy Agency (IRENA), the region's investment in renewable energy excluding traditional hydropower has grown 11-fold (IRENA 2016). To put this in perspective, the global growth in investments in renewable energy during the same period was only sixfold. Countries in the region are increasingly gravitating away from conventional energy sources. In 2015, Mexico and Chile joined Brazil on the list of the ten foremost markets in renewable energy in the world. Other countries in region have demonstrated a willingness to develop and/or implement innovative policy instruments. Three such policy instruments, renewable power auctions, solar thermal requirements, and biofuels blending mandates are already in use in the region. They have been responsible for significantly reducing the cost of renewable energy in LAC.

The list of the region's accomplishments in the renewable energy arena is considerably long; however, the following specific achievements emerge as the most prominent (GRC 2017). First, Chile launched its maiden geothermal energy plant. This, incidentally, was also the first of its kind in the entire Latin America and Caribbean region. At installation, the plant was billed to provide electricity to 165,000 homes. It boasts several positive attributes. Apart from the power it provides to its clients, the plant helps Chile to avoid 166,000 tons of carbon emissions. Second, Brazil, a regional leader in renewable energy operates the two largest solar parks in the region. The two Brazilian solar parks with about 1.8 million solar panels, supply clean energy to 600,000 homes in the country. Third, Guatemala produces 71% of energy from renewable sources. Fourth, Honduras produces 61% from renewable sources. Fifth, Uruguay produces 62% of its energy from renewables. All told, the Latin America and Caribbean region is at the forefront in efforts to promote the use of renewable energy. The region made many noteworthy achievements in the energy domain in 2017 (GRC 2017). Brazil, Chile, Uruguay and Mexico rank among the ten best countries on the Bloomberg New Energy Leaders in 2007. The region's investments in the renewable energy industry increased by 65% in 2017 to US$17.2 million (Sweigart Online). This surpassed the global average increase of 3%. Record investments occurred in Mexico. In Argentina, investments in renewable energy increased ninefold. Other countries of the region that have also been active in the renewable sector in the recent past include, Brazil, Peru, Chile, Costa Rica, the Dominican Republic, and El Salvador. The renewable energy source of choice for most of the investors are solar and wind.

6.8 Water in Built Space

The water sector in Latin America and the Caribbean (LAC) has undergone significant institutional reforms since the 1960s. However, the most significant reform measures did not occur until the 1990s (Foster 2005; World Water Council Online). This is when most countries in the region embarked on activities to reform their water supply sector. Chile led the way in these efforts. In this regard, Chile passed a new piece of legislation designed to liberalize its water sector in 1988. Argentina and Mexico followed in Chile's footsteps 3 years later in 1991 by authorizing private sector participation in the water provisioning process. Peru, Colombia, and Bolivia later joined the list in the 1990s by liberalizing their water sector. Brazil and the rest of Central America soon followed. By the end of the 1990s, only a few countries in the region were yet to reform their water sector. Prior to the 1990s, many Latin American governments such as Argentina, Chile, Colombia, and Peru operated water sectors that were, for all practical purposes, national monopolies directly controlled by their national governments (Foster 2005: 3).

The changes occasioned by the reforms of the 1990s ushered in a new era in the life of water provisioning in LAC. Currently, municipalities play important roles in the water domain. In many cases, these entities are directly responsible for water provisioning. Examples of these cases include Brazil Colombia, Ecuador, Guatemala, Mexico, and Peru. The case of Colombia and Mexico is particularly noteworthy because this responsibility was until the 1980s in the hands of national service providers. Currently, and especially subsequent to the decentralization and reform exercises of the 1980s, special units have been created within municipal governments to specifically take care of water and other utility provisioning. The units typically have a budget quite apart from that of the municipal government of which they are a part. National water and sewer companies, some of which were created in the 1960s and 1970 remain in operation to date in Costa Rica, Dominican Republic, El Salvador, Haiti, Panama, Paraguay, and Uruguay. Thus, the reforms notwithstanding, public utility entities continue to be active in the water domain in Latin America. By some accounts, 90% of urban water and sanitation services in the region are provided by such entities (Foster 2005). An important factor accounting for this phenomenon is the fact that many of the concession contracts that were signed in the 1990s were either later renegotiated or entirely annulled. Prominent in this regard is the concession for the Buenos Aires-based Aguas Argentina and that for Cochabamba and La Paz both of which were based in Bolivia. Water services in most rural areas in the region are provided by community organizations (i.e., Juntas de Agua). Funds for building such infrastructure are usually provided by the national governments while the community typically contributes the labor input.

The water sector in LAC has registered considerable success; the proportion of people with access to water has risen from 33% in 1960 to 85% in 2000 (World Water Council Online). However, as many as 77 million people in the region remain without access to potable water. Of these, 51 million are based in rural areas, while 26 million live in cities. Thus, considerable disparities characterize the region's water sector. The most conspicuous of these inequalities exist in terms of the water cost burden. In this regard, the poor pay as much as 1.5–2.8 times more than non-poor for water (World Water Council Online). Although this connotes financial constraints at the individual level, it is worth noting that municipal and other authorities responsible for water provisioning in the region face different degrees of financial problems. These problems are compounded by natural factors having to do mainly with water scarcity.

Most of the region depends on aquifers for water. However, many of the aquifers in the region are facing increasing threats from several factors. Prominent in this connection are overexploitation and pollution. For example, as many as 102 of the 653 aquifers are overexploited. The main sources of pollution in the region are mining and agriculture. The shortage of usable water occasioned by pollution, overuse and other factors have caused people to reconsider their vocations and avocations. For instance, people are increasingly abandoning more profitable but water-intensive agricultural activities such as cotton-farming; instead, they are taking up less lucrative and less water-intensive agricultural activities such as grain crop farming.

Aquifers are not the only sources of water under threat in the region. Lakes and rivers are also increasingly being contaminated by pollutants from different sources, including agriculture and industries. Also worth noting are the devastating effects of hurricanes not only on human life but also on water resources in the region. Populations in hurricane paths have been known to bear the brunt of this catastrophic natural disaster. People have been killed, displaced and rendered homeless on many occasions by this event. For instance, in 1998, Hurricane Mitch claimed 9000 lives in Central America alone. In addition, it caused damages amounting to 6 billion US dollars while displacing 75% of Hondurans. As for its direct impact on water resources, the effect of hurricanes is paradoxical; they have been known to cause both floods and droughts in the region.

An important but oft-ignored attribute of water as an important resource for human life is its transboundary nature. This attribute may have local, national and international dimensions. In the case of LAC, there are many water basins in the region that straddle international boundaries. Some of these straddle regional boundaries such as the sub-region between Mexico and the United States. In this case, decades of agricultural and industrial run-off from intensely chemicalized farms in the United States discharging into the ground have contaminated ground water sources and aquatic life in Mexico and beyond. In some cases, the question may be fairness with respect to use given the possibility that one country may be inclined to use more than its own fair share of a transboundary water resource.

Mitigating against disagreements and conflicts arising from such situations necessitates stakeholders in transboundary water and other resources entering into signed agreements on their rights over such resources. One such agreement—that over the Plata Basin—involving Brazil, Argentina, Paraguay, and Uruguay comes to mind.

6.9 Concluding Remarks

Few built spaces in the world reflect the cultural and political history of the wider region in which they are located as those of Latin America and the Caribbean (LAC). Here, nature in built space is a reflection of two important historical factors, namely the region's indigenous culture and colonial history. These factors significantly affect efforts to address the food, energy and water needs in built space throughout the region. In fact, the domestication of plants predated the arrival of Christopher Columbus in the region in 1492. Evidence of the use of hydroponic techniques that enable the growing of crops in water exists in the region's pre-colonial history. Meanwhile, efforts to promote urban greenery constituted part of the urban planning schemes that were superimposed on the region's spatial landscape during the European colonial era. The introduction of the Garden City Model, a brainchild of an Englishman by name, Ebenezer Howard, exemplified this trend in the region. This history has been considerably successful in sustaining initiatives to promote nature in built space in the region. However, since the onset of the twentieth century, these initiatives have faced enormous challenges from urbanization and commensurate forces.

The region's rapid rates of urbanization have, since the twentieth century, presented enormous challenges to the initiatives. The region's heightened rates of urbanization continue unabated. During the 2005–2015 decade, 15 cities in the region transitioned from the 'small' to the 'medium size' category; meanwhile two cities in the region evolved to become the world's newest megacities. This suggests that high levels of urbanization promise to remain a permanent feature of the LAC region. Therefore, only those initiatives to promote nature in built space that are cognizant of the region's high levels of urbanization have any chance of succeeding now and in the future.

References

Atlantic Council (Online) Urbanization in Latin America. Accessed 26 Aug 2019 at: http://www.atlanticcouncil.org

Bergvall M, Dahl E (2015) A Green Belt of synergies – a study on the implementation of a contemporary Green Belt, A Masters Thesis. Landscape Architect Programme, Upsala

BTI (Online) Latin America and the Caribbean: Growing Frustration, Lagging Response. Bertelsmann Stiftung Transformation Index. Accessed, October 23, 2019 via: https://www.bti-project.org/en/home/

Dussan M (1996) Electric power sector reform in Latin America and the Caribbean. Inter-American Development Bank. Working Papers Series IFM-104

ECLAC (2019) Poverty in Latin America remained steady in 2017, but extreme poverty increased to the highest level since 2008, while inequality has fallen notably since 2000. Press Release by the Economic Commission for Latin America and the Caribbean. January 29, 2019. Accessed 27 Aug 2019 at: https://www.cepal.org/en/pressreleases/poverty-latin-america-remained-steady-2017-extreme-poverty-increased-highest-level

Eyre LA (1990) The tropical national parks of the Latin America and the Caribbean: present problems and future potential. Yearbook (Conference of Latin Americanist Geographers) 16: 15–33.

FAO (2014) Growing greener in Latin America and the Caribbean: an FAO report on urban and peri-urban agriculture in the region. Food and Agriculture Organization, Rome

FAO (2019) Urban and peri-urban agriculture in Latin America and the Caribbean: overview. Accessed 28 Aug 2019 via: fao.org/ag/agp/greenercities/GGCLAC/overview/html

Foster V (2005) Ten years of water service reform in Latin America: toward an Anglo-French model. Paper for the World Bank. Water Supply and Sanitation Sector Board Discussion Paper Series No. 3

GRC (2017) Latin America 2017 top three green trends. A Posting by the Geothermal Resources Council (GRC). Accessed 29 Aug 2019 at: https://www.flickr.com/photos/geothermalresourcescouncil/36406847560/in/...

Huber E, Nielsen F, Pribble J, Stephens JD (2006) Politics and inequality in Latin America and the Caribbean. Am Sociol Rev 71(6):943–963

IPS (2019) Devastating epidemic of crime & insecurity in Latin America and the Caribbean. An Inter Press Service (IPS) publication. Accessed 28 Aug 2019 via: http://www.ipsnews.net/2019/05/devastating-epidemic-crime-insecurity-latin-america-caribbean/

IRENA (2016) Renewable energy market analysis: Latin America. In: International Renewable Energy Agency (IRENA). IRENA, Abu Dhabi

Jaeger T (2013) Natural world heritage in Latin America and the Caribbean: options to promote and underutilized conservation instrument. An independent review prepared for IUCN

Mycoo M (2006) The retreat of the upper and middle classes to gated communities in the post-structural adjustment era: the case of Trinidad. Environ Plan A: Econ Space 38(1):131–148

Netroots (2019) Religion and Indians a century ago, 1919. Accessed, 4 Nov 2019 via: http://nativeamericannetroots.net/diary/Indians-101-Religion-and-Indians-a-Century-Ago-1919

New World Encyclopedia (2019) Spiritualism. Accessed, 25 Sept 2019 via: https://www.newworldencyclopedia.org/entry/Spiritualism

Njoh AJ (2015) French Urbanism in Foreign Lands. Springer, New York

Olupona, JK (2015) The spirituality of Africa: interview by Anthony Chiorazzi. Harvard Gazette. Accessed on 14 Oct 2019 via: https://news.harvard.edu/gazette/story/2015/10/the-spirituality-of-africa/

Schmid R (2007) Squashes show ancient farming in South America. The Guardian (Fri 29 Jun 2007). Accessed 26 Aug 2019 at: https://www.theguardian.com/world/2007/jun/29/international.mainsection

Sweigart E (Online) Just the numbers: investments in renewable energy sources is booming in Latin America. Accessed 3 Sept 2019 via: americasquarterly.org/content/investment-renewable-energy-sources-booming-latin-america

Terraza H (Online) Making Latin American cities better: urban development. Accessed 29 Aug 2019 via: https://www.americasquarterly.org/content/expert-commentary-urban-development

The Conversation (2019) What makes a mountain, hill or prairie a 'sacred' place for Native Americans? Accessed, 3 Oct 2019 via: https://theconversation.com/what-makes-a-mountain-hill-or-prairie-a-sacred-place-for-native-americans-73169

UNEP (2016) Global Environmental Outlook (GEO-6): Regional Assessment for Latin America and the Caribbean. United Nations Environmental Programme. Retrieved, August 25, 2019 at: http://www.unep.org/publications

UNEP (Online) The state of biodiversity in Latin America and the Caribbean: a mid-term review of progress towards the AICHI biodiversity targets. Accessed online, 10 Oct 2019 via: https://www.cbd.int/gbo/gbo4/outlook-grulac-en.pdf

UN-Habitat (2009) Revisiting urban planning in Latin America and the Caribbean. Report prepared for the UN-Habitat by C. Irazabal. United Nations Centre for Human Settlements, Nairobi

Vasquez A, Devoto C, Giamoti E, Velasquez P (2016) Infrastructure systems facing fragmented cities in Latin America – case of Santiago, Chile. Procedia Eng 161:1410–1416

Vergara W, Isbell P, Rios AR, Gomez JR, Alves L (2014) Societal benefits from renewable energy in Latin America and the Caribbean. An online publication of the Inter-American Development Bank (IDB). Accessed 3 Nov 2019 via: https://publications.iadb.org/en/societal-benefits-renewable-energylatin-america-and-caribbean

World Water Council (Online) Water problems in Latin America. Accessed 03 Sept 2019 via: forest-trends.org/wp-content/uploads/valorandonaturaleza/water_problems_22.03-04.pdf

WWF (2019) Impacts of climate change in the Amazon. A World Wildlife Fund posting. Accessed 28 Aug 2019 at: https://wwf.panda.org/knowledge_hub/where_we_work/amazon/amazon_threats/climate_change_amazon/amazon_climate_change_impacts/

Chapter 7
Nature in Built Space in North America

Abstract North America, comprising the United States and Canada, is the least socio-politically diverse, but one of the most geographically extensive of United Nations Environmental Programme (UNEP) regions. It ranks as the most economically and industrially developed. These unique features have far-reaching implications for nature in the region's built space, particularly its urban areas. This chapter discusses the ways in which these features conspire with the region's politico-economic, socio-cultural geo-ecological and historical factors to facilitate or obfuscate efforts to promote nature in the built environment. It is shown that the region's liberal democratic and economically advanced system, combined with its dominant Judeo-Christian culture, have proved to be a boon for these efforts.

7.1 Introduction

The United Nations Environmental Programme (UNEP) groups the United States and Canada under one category, namely North America. The region falls under the monitoring authority of the Regional Office for North America (RONA). Shown as Fig. 7.1, this region groups together two of the three largest countries in the world. The present chapter focuses on this region. In particular, the chapter analyzes nature in built space as a function of the region's political, economic, social, technological, ecological, cultural and historical (PESTECH) factors. The region is bordered to the south by Mexico, to the West, by the Pacific Ocean; the Arctic Ocean marks the region's northern border while the Atlantic Ocean marks its eastern frontier. It is the most economically progressive region in the world. A historical factor worthy of note with respect to the region is the fact that it was colonized by European powers, particularly, Spain, Britain and France. Consequently, its cultural patterns possess European roots.

Apart from a stellar economic record, the North America region is also on record for registering significant improvements in its environmental situation especially within the last decade. These improvements, according to the United Nations Environmental Programme (UNEP), have largely been due to investments in meaningful institutional efforts (UNEP 2016: 6). These have consistently comprised well thought-out investments, data collection and analysis initiatives. These have yielded remarkable dividends as the region's environment and land have typically been

A. J. Njoh, *Nature in the Built Environment*,
https://doi.org/10.1007/978-3-030-39759-3_7

Fig. 7.1 New York Central Park, a testament to the fact that an urban forest is possible even in one of the world's most densely populated and modernized cities. (Source: Wikimedia.com. Available free of charge from: https://commons.wikimedia.org/wiki/File:Central_Park_(from_Parker_New_York).jpg)

judged as being in 'good shape' (UNEP 2016: 6). North America does not face any problem of land shortage despite the fact that certain sub-regions, such as the Northeast, Southeast and Southwest of the United States, and the Southeast of Canada, are densely populated. However, land throughout the region is under threat from soil pollutants resulting from agricultural and industrial activities. Concern with the health of land in the region has both environmental and economic development rationales. In the US alone, more than two million people are employed in forestry and agriculture. Canada's GDP for agriculture, forestry, fishing and hunting was US$24.7 billion in 2010. As for pollution, North America contributes to the global air pollution problematic directly and indirectly. The region's excessive reliance on automobile transportation contributes directly to the 'smog' and traffic congestion problems of major cities in the region. As the world's leading importers of processed goods, the region also contributes indirectly to the global air pollution problematic. For instance, 8–12% of China's CO_2 emissions were attributed to exports to the US alone between 2002 and 2007. Finally, there is the problem of biodiversity loss; this cannot be reversed without serious efforts to promote nature in urban space. Despite its vast expanse of land and sea, the region has suffered significant losses in its biodiversity as a result of population growth, urbanization, and large-scale agriculture. Commensurate with this has been a loss, estimated to be about 28% and 27% respectively in grassland and dryland species (UNEP 2016). This chapter delves into the nature of the major environmental problems with implications for built space that characterize the North America region. Then, it identifies and discusses the major politico-economic, socio-cultural geo-ecological factors facilitating or obfuscating efforts to eradicate the problems.

The chapter progresses in the following order. The next section presents important background information emphasizing politico-economic, socio-cultural, geo-ecological and technological profile of the region. Next, it discusses an important cultural element of the region, namely religion, and its implication for environmental stewardship. Then, it analyzes initiatives to develop and maintain forests, farms and gardens, energy and water resources in built space in the region.

7.2 North America Through a PESTEH Lense

The North America region comprises two countries, the United States and Canada that rank among the most democratic in the world. An important dimension of democracy as practiced in the region is the autonomy of municipal governments. This is especially critical in the context of the discussion in this book because of the administrative and regulatory role of municipal governments in the creation of nature in built space. Here, it is necessary to note that comprehensive national government urban planning is neither an American nor a Canadian tradition thanks to an anti-government ideology that prevails in both countries.

Geo-ecologically, the region is considerably vast and diverse. Consider the case of the United States; there is no consensus on the specific number of sub-environmental regions comprising the entire country. The United Nations Environmental Programme (UNEP) contends that the country can be divided into ten or more environmental sub-regions (UNEP 2016). For simplicity sake, we adopt the UNEP 10-sub-region schema, which produces the following sub-regions: the Artic, Boreal, Forest, Cascadia, Great Basin, California, Samora, Prairie, Laurentia, and Dixon. The Artic is a vast territorial band stretching from the Pacific Ocean on the West to the Atlantic in the East. On its part, the Boreal constitutes a second band adjoining the Arctic from the Pacific to the middle of the continent while the Forest sub-region continues from the Boreal to the Atlantic Ocean. Three sub-regions, Cascadia, Great Basin, California, and Samora, line up (in that order) from top-to-bottom along the Pacific Ocean and extends to about a third of the US mainland. The Prairie stretches from the bottom of the Boreal all the way down to the Gulf of Mexico. The Laurentia lies atop Dixon and stretches from the Prairie to the Atlantic Ocean.

The more common practice in socio-economic and politico-administrative discourse is to group the country into four main regions, including the Northeast, South, North Central, and West. The Northeast is densely populated and has a history of industrial production. This has far-reaching implications for nature in built space in the region and beyond. Prominent among the problems often associated with the region's history of industrial production and high population density are acid rain damage resulting from coal-burning power plants, and air pollution resulting from a heavy concentration of automobiles. The South is characterized by heavily populated urban areas interspersed among sparsely populated rural areas. The population has historically been employed in agriculture, textile, paper mill and chemical industries. Consequently, soil pollution from fertilizer and pesticide use as well as noise pollution from industrial plants pose significant challenges to efforts to promote nature in built space in the sub-region.

The North-Central sub-region is, like the South, characterized by heavily populated urban centres punctuated by sparsely populated rural areas. It also lays claim to a history of agriculture and heavy industry, aging factories and power plants. This explains its nickname, the Rustbelt—a nickname it shares with the Northeast, which also boasts a plethora of aging factories and power plants. Built space in the area has, over the years, been confronted with problems of soil pollution from heavy fertilizer and pesticide use, noise pollution from factories and power plants, and water contamination from lead, pesticides, fertilizers and other contaminants that find their way into ground water sources.

The West is comprised of large arid areas in the hinterland and forests in the heavily populated areas along the Pacific Ocean. It is sparsely populated in the hinterland areas. On average, the entire sub-region has been experiencing population growth in the recent past. It has a rich history of ranching, agriculture, mining and oil production. These features conspire to threaten the sustainability of nature in built space. The most densely populated areas, particularly California face water scarcity problems that are compounded by sporadic wildfires that have become more frequent in recent times with the rapid growth of the population. Soil pollution resulting from fertilizer and pesticide use is a problem in the hinterland areas where agriculture constitutes the mainstay. Air pollution—and in fact, smog—from the heavy concentration of automobiles is the leading problem affecting built space in southern California.

Socio-economically, the U.S. and Canada have experienced four significant developments with implications for the creation and maintenance nature in built space during the last century. The region has been one of the most economically progressive in the world since World War II. The economic systems of the two countries comprising this region are essentially alike. Both are market-oriented and have evolved from being largely agricultural to their current form, which is primarily industrial and urban. The two countries are endowed with natural resources such as oil, natural gas, forests and aquatic products. Thus, the need to protect nature in built space assumes a heightened level of importance in the rapidly industrializing and rapidly declining size of the natural environment in the region. One favorable factor at in this region is its highly successful economic profile. The current economic development profiles of the two countries are very telling in this regard (Mundi Index Online). The 2017 estimated gross domestic product per capita (GDP per Capita) for Canada and the United States are respectively $48,100 and $59,500; their median incomes are $45,220 and $60,336. Unemployment is low in both countries, with Canada and the United States reporting rates of 6.0% and 4.0% respectively in 2017. These statistics have implications for efforts to create and maintain nature in built space to the extent that such efforts require significant material and financial resources.

The second development with implications for nature in built space is of the demographic genre. This mainly has to do with so-called 'Millennials" as a proportion of the total population in North America. Millennials, that is, persons aged 23–38, are fast emerging as the largest adult population in the region. In the United States, according to the Pew Research Center (PRC 2019), members of this age cohort are on track to surpass the population of 'baby boomers' or persons aged 55–73 by the end of 2019. The US Census Bureau reported that the population of baby boomers stood at 74 million while millennials numbered 71 million in 2016. By the end of 2019, the population of millennials in the United States rose to 73 million; in contrast, the number of baby boomers dropped to 71 million. One implication of this demographic trend is an increase in the demand for nature, especially recreational facilities such as playgrounds in built space. This is because the obvious rise in the most active segments of the population—young adults and their offspring.

The Pew Research Center drew attention to three more demographic trends with the potential of significantly affecting the demand and supply of nature in built space in North America. The first of this is the projected increase of the Hispanic population in the region. In the United States, Hispanics are projected to surpass African Americans and become the largest minority group in the country. More precisely, Hispanic population is on track to reach 32 million in comparison to 30 million African Americans by the end of 2019. Given that Hispanics are devoutly religious and mainly adherents of Catholicism, it is safe to expect a stronger influence on environmental stewardship rooted in religiosity. The second observed demographic trend is a significant increase in the number of unmarried parents. This population as a proportion of the total has been on the rise since 1968. It grew by more than 13% in 1968 to 32% in 2017 (Pew Research Center 2019). The third observed trend is an increase in the population of foreign-born persons in the region's population. This increase is occurring in both the United States and Canada although in the former case, there has been a slight decline as a proportion of the total population. Nevertheless, the population has increased significantly especially within the last decade. Two facts are worth underscoring as far as this population goes. First, most members of this population come from resource-scarce countries. This suggests that most immigrants in North America know a thing or two about doing more with little when it comes to the use of resources. It is therefore reasonable to expect an intensification of urban greening efforts to the extent that a primary objective of such efforts is to maximize the utility of space. Second, immigrants, especially the newly arriving ones, almost always prefer urban as opposed to rural locations. Therefore, one unavoidable upshot of increases in immigrant populations is urban expansion in both demographic and physical terms. This invariably increases the demand for green infrastructure, especially farms and food trees that can be effective in supplementing urban food supply systems.

A fourth development with implications for nature in built space in North America is industrialization. This began with the Industrial Revolution (IR), whose roots are traceable to England of the eighteenth century. One of the best-known impacts of IR in North America was on rural areas. To appreciate this, one must first understand that IR, among other things, mechanized agriculture and manufacturing. This effectively provided the impetus for freed farmhands to migrate from rural areas to take up jobs in manufacturing plants in urban areas. However, the growth and development of these areas was constrained by the limitations of pedestrian and technologically rudimentary transit systems. All of this changed with the invention of the automobile, or what has been labelled the most significant agent of change in North America (Encyclopedia Britannica 2019). The automobile engendered the remarkable growth and proliferation of urban areas that has occurred in the region since the beginning of the twentieth century. It is also incriminated for some of the most atrocious assault ever visited upon the natural environment in the region. Thus, no discussion of nature in built space in North America can be deemed complete without mention of how it has been, and is being, shaped by the automobile. After all, the automobile has demonstrated a voracious appetite for land, and is a leading cause of air pollution in built space. Much land in the region is devoted to road transportation; this includes land dedicated to roads, road rights of way, streets and parking lots alone. By some estimates, as much as 30–60% of urban space is consumed by transportation infrastructure in North America; this proportion rises significantly in some areas such as California, where it is estimated to be 70% (Rodrigue 2017).

The final development or factor with implications for nature in built space discussed here relates to urbanization. North America has experienced very high levels of urbanization since it emerged from the yoke of European colonialism in the eighteenth century. This factor is particularly important because of its direct impact on the natural environment. For a long time during the region's modern history, the development of any built space literally meant the complete replacement of the natural features of that space with man-made objects such as buildings and streets. It also entailed the leveling of hills, and the filling of valleys and wetlands. However, with increased environmental awareness, this changed. The change was marked by the release of Ian McHarg's classic, *Design with Nature*, which essentially underscored the need for human settlement development efforts to avoid or at least minimize any alteration to the natural landscape (see McHarg 1971).

A renowned landscape architect, McHarg sought to demonstrate the unavoidable and inextricable link between natural and urban environments. Contrary to the prevailing opinion, he believed that the relationship between the two environments could be harmonious as opposed to conflictual. However, harmony could only be assured by summoning the most innovative spatial management strategies. The need for such strategies is accentuated multiple folds today by the rapid demographic and physical expansion of built space. To be sustainable, urban planning must acknowledge the symmetric relationship between nature and built space. Thus, built space shapes, and can be shaped by, nature. Influenced by the extraordinary wealth of their nations, North Americans have ranked among those most likely to test the limits of both environments' tolerance for each other. In this regard as Tarr (Online) has observed, North Americans are well-known for seeking to live in coastal areas, along rivers, hilltops, and similar locations where nature offers limitless attractions. Yet, these areas have served as the stage whereupon nature has demonstrated its most devasting and ferocious vigor. Built spaces that have been shattered by floods, hurricanes, and tornadoes constitute indelible testaments to this phenomenon in North America.

The paradox is that North America counts among the regions with the most restrictive environmental regulations in the world. This, of course, was until the emergence of the Trump Administration and the commensurate efforts to undo almost all Obama era environmental regulations in the United States. Nevertheless, the region is home to leading institutions of higher education and research on the natural environment. It is also a leading source of technical and financial support for environmental initiatives worldwide. Thus, it is safe to conclude that the region maintains its place as a leading actor in efforts to promote the harmonious co-existence of the natural and built environments. The roots of such co-existence run deep in the region's history. The following account by Tarr (Online) is quite telling in this respect.

> In the 19th century, for instance, the demands of city dwellers for food produced rings of garden farms around cities and drove the transformation of distant prairies into cattle ranches and wheat farms; and the many horses quartered in cities required feed, consumed the products produced by thousands of acres.

To be successful, therefore, contemporary efforts to promote nature in built space in North America would do well to draw from lessons of experience in the region's rich history and culture. A dominant element of this latter is religion. As stated earlier in this chapter, Christianity is the dominant religion in the region. Consequently, it is in order to review major features of the Christian doctrine with a view to understanding how it promotes or discourages environmental stewardship.

7.3 Christianity and Environmental Stewardship

Christianity has the largest number of adherents (2.1 billion) of all religions in the world; it is also globally the most widespread. It comprised a major component of the cultural goods that were introduced in foreign lands by European powers during the heydays of colonialism. It was a crucial element in efforts on the part of Europeans to discharge their professed preordained mission to civilize 'cultural others.' Thus, the Christian doctrine as articulated in the Holy Bible has been influencing belief systems in many societies since Europeans started venturing into other parts of the world about 800 years ago. In what many consider a groundbreaking piece on the religious roots of contemporary ecologic crises, Lynn White (1967) posed the following perennial and thought-provoking question. "What did Christianity tell people about their relations with the environment"? White begins his attempt to answer this question by recounting the Greco-Roman version of the creation myth, which he characterizes as incoherent. The incoherence is probably a function of the fact that ancient intellectuals such as Aristotle did not subscribe to any theory that posits a beginning for the visible world. In fact, Aristotle and other intellectuals of that era harboured a theory that viewed time as cyclical and repetitive (White 1967). Within this framework, the world has neither a beginning nor an end.

However, the question of how Christianity instructs people to handle or relate to nature cannot be adequately answered without consulting the Bible. Here, we note the diametrically opposed way in which the Christian view of time and the world differs from that conveyed in Greco-Roman mythology. In contrast to this latter, Christianity impresses upon its adherents that time is nonrepetitive and linear, and that the world is God's creation. Thus, the world has a definite beginning, and every creature therein has an assigned role. Within this framework, as stated in Genesis 1: 26–28, man was afforded dominion over all other creatures. This passage as gleaned from the King James Version reads as follows:

> And God said, Let us make man in our image, after our likeness: and let them have dominion over the fish of the sea, and over the fowl of the air, and over the cattle, and over all the earth, and over every creeping thing that creepeth upon the earth. So God created man in his own image, in the image of God created he him; male and female created he them. And God blessed them, and God said unto them, Be fruitful, and multiply, and replenish the earth, and subdue it: and have dominion over the fish of the sea, and over the fowl of the air, and over every living thing that moveth upon the earth.

As noted earlier, Christian theologians have been ferocious in countering the charge that Christianity emerges as the most anthropocentric religion ever known to mankind. If these theologians' task in this regard appears analogical to an uphill battle, it is because of the contents of several Biblical passages. Apart from the afore-quoted, the Bible is unequivocal in presenting humans as the creatures with dominion over all other creatures. In concert with this pronouncement, the Bible grants humans the right to "subdue" everything on Earth. White (1967) persuasively argues that the desire to 'subdue' the Earth and all its creatures constitutes the motivation for Western science, whose global reach and dominance is irrefutable. To be sure, this is by no means intended to trivialize the contributions of other civilizations and cultures such as the Chinese, Japanese, Islamic scientists of the Middle Ages. Yet, there is no denying, that Judeo-Christianity teleology has been the trendsetter throughout the world for more than a millennium. In fact, as White (1967: 1205) observes, "we continue today to live as we have lived for 1700 years, very largely in a context of Christian axioms." How did this come to be? In other words, how did Christianity go about supplanting other religions and imposing the Judeo-Christian view of nature on the world at large?

The missionaries worked in tandem with European colonial authorities who had two main overlapping goals: economic exploitation and acculturation in non-settler colonies; and political, social and cultural domination in colonies of the settler variant. They encountered fierce resistance in both cases but the forces of resistance could not, like the colonial invaders or aggressors possess the military capacity to meaningfully defend their territories. Consequently, the aggressors succeeded at not only at forcefully acquiring physical territory but also at supplanting the indigenous and other non-Christian religions with Christianity. Additionally, colonial authorities and Christian missionaries created and strived to address a need non-Western and non-Judeo-Christian societies did not even know existed. This is the need to earn eternal salvation and to save oneself from eternal punishment. They preached that this need can be met if and only if one accepts and worships exclusively one God, the 'Christian God.' When it came to supplanting the indigenous belief systems and converting non-Westerners to Christianity, they retooled the tactics that had been employed by their predecessors, Early Christians in ancient Rome.

In his instructive, if only controversial, book on this subject Bart Ehrman (2018) contends that Early Christians employed five tactics to convert pagans to Christianity in ancient Rome. These included, creating a need and professing to meet it, asserting the presumed superiority of Christianity, working from the bottom-up, eliminating its competition, and enlisting powerful supporters. Ironically, the same tactics were effectively employed in efforts to globalize Christianity and institute Judeo-Christian notions of environmentalism several centuries later. Although these notions were packaged and delivered as Christian doctrines, they were, by design or otherwise, richly laced with elements of capitalism. For instance, there was a conscious effort to accentuate the economic as opposed to the sacred value of natural resources. In this connection, missionaries moved speedily to strip rivers, forests, mountains, and other natural features of the sacred status they enjoyed in foreign lands prior to the European conquest. One consequence of this was the loss of the reverence for nature that served as a deterrence to environmental abuse. We shall return to this subject later on in this chapter. For now, it is informative to cursorily examine the various strategies employed by Christian missionaries to supplant other belief systems in the non-Western world.

Asserting the Presumed Superiority of Christianity The second tactic involved Christian missionaries working fervently to assert the 'Christian's God' as the most powerful. This was premised on the notion that divinity is all about power. Consequently, the missionaries, and Christianity as a whole, waged a ferocious war against other faiths and forcefully promoted the idea that their own God was the most powerful. He is capable of better executing all the duties of the 'indigenous gods'—healing the sick, preventing natural and other types of disasters, increasing crop yields and so on. Christian narratives are replete with assertions designed to demonstrate the superiority of Christianity over other religions or belief systems. Ehrman (2018) recounts one such narrative culled from the Acts of John, an apocryphal or noncanonical piece of Christian writing. The narrative talks of the missionary travails of one of Jesus Christ's disciples, John, the Son of Zebedee. John is said to have visited the temple of the goddess Athena in the city of Ephesus. Upon entering this sacred site wherein a large crowd of pagans had gathered, John challenged the crowd to pray to their god to strike him dead; should the god fail to respond, he will ask his God, the 'Christian God' to kill them. Recalling that John has accomplished many feats such as resurrecting dead people, the crowd was terrified by his proposition. Noting the crowd's refusal to take his challenge, John proceeded to curse the sacred place's divinity; this resulted in the altar of Artemis splitting into pieces, while the idols broke apart and the roof caved in killing the chief priest instantly. Deducing from this miraculous act that John's God, the 'Christian God,' is the most powerful and shouted in unison, 'there is only one God!' Christians have been employing miracles as an evangelistic calling card since time immemorial. The many miracles of Jesus Christ, including the healing of Peter's mother in-law (see Matthew 8: 14–15, Mark 1: 29–31, Luke 4: 38–41); resurrecting Lazarus of Bethany (see John 11: 1–44); and walking on water (reported in the gospels of Matthew, Mark and John). Christian missionaries that participated in the European colonization of foreign lands paraded these and other miracles to bolster the case of the 'Christian God' as the most powerful. One reason why this might have been well received in non-Western societies is the fact that Christian doctrine elevated humans above all other creatures—until then, an alien notion in those societies.

Working from the Bottom-up Christian missionary activities bent on spreading Christianity to all the nooks and crannies of the world, understood, rather quickly, the importance of working from the bottom- or grassroots-up. Working from the bottom-up is a tactic that entailed directing conversion initiatives at grassroots or the lower crusts of society. By doing so, Christian missionaries avoided confronting members of the most powerful echelons of society; these are usually the most devout adherents to the society's indigenous norms and beliefs. In Africa, for instance, Christian missionaries directed their conversion efforts mostly at women as they were considered the path of least resistance. Here, the application of Western ethos disguised as Christian doctrine discouraging polygamy is very interesting. Christian missionaries excluded men but not women involved in polygamous relations from full-fledged membership in the church. This was an upshot of the missionary's knowledge of the influence that women wielded over children during their formative years in Africa. Therefore, to convert women is to convert the children. Also, African women are known to be more malleable in the cultural milieu than their male counterparts. Directing conversion initiatives at the grassroots also meant targeting the poor and destitute. Members of this group who are always plentiful, in Africa, are gullible and therefore easily convertible. Christian missionaries found several passages in the Bible to cull from in efforts to convert members of this most vulnerable group. In the gospel of St. Luke (6: 20), it is clearly stated that, "Blessed are the poor." In an earlier book of the New Testament, Jesus begins his Sermon on the Mount, by stating that "Blessed are the poor in spirit for theirs is the Kingdom of Heaven" (Matthew 3: 5, NKJV).

Eliminating Its Competitors Yet another tactic typically employed by Christian missionary in non-Western lands is the elimination of the competition. Missionaries required those who converted to Christianity to disavow their previous and other faiths. Conversion was supposed to be a final and irreversible decision. Converts were not allowed to swing between Christianity and other religions. Rather, they were expected to commit exclusive allegiance to Christianity. The missionaries maintained that Christianity was the only way to heaven. Therefore, anyone who elected to commit to any religion other than Christianity was bound for hell. Given the many people who converted to Christianity during the European colonial era in the Americas and Africa, it is safe to state that the missionaries were exceedingly successful. It is important to note that the gains for Christianity translated directly into losses for indigenous religions. Missionaries went so far as to discourage Christians from fraternizing with adherents to other religion. While missionaries did not adopt violent means such as what obtained as part of the Crusades or religious wars that the Latin Church sanctioned in the medieval times, they missed no opportunity to show their disdain for non-Christians.

Winning Over Powerful Supporters Christian missionaries were not the first foreign religious authorities to arrive many parts of the world outside of Western Europe, the citadel of Christianity since the 2nd century. In Africa and parts of Asia, for instance, Islam had preceded the arrival of Christian missionaries by a millennium. In addition, these areas have a rich history of vibrant indigenous religions. Consequently, Christianity faced fierce competition, and in some cases, outright rejection. In such instances, it was not sufficient to win over women as a strategy for winning the children, hence, the majority of the population in any given community. Rather, the success of the Christian missionaries depended largely on the extent to which they could persuade village chiefs, clan heads and other local leaders in foreign lands. Winning the hearts and minds of these leaders was necessary to gain authorization to operate in the communities under their politico-administrative control.

The dominance of Christianity and the Christian ethos throughout most of the world, especially regions that experienced European colonialism constitutes irrefutable testament to the resounding success enjoyed by the conversion strategies enumerated above. Thus, it is safe to state that the way people, communities and governments throughout most of the world view and treat nature today is influenced by Christian doctrine and ethos. These, in turn, are a corollary of Christianity's desire to attain other, sometimes disparate, objectives. Paradoxically, foremost amongst these is the economic success of its adherents. Efforts to attain this objective, as Lynn White (1967) argues, account for the fact that capitalism is the favored ideology in most countries where Christianity is the dominant religion. The efforts are also at the root of contemporary environmental crises.

7.4 Urban Forests and Green Space

7.4.1 Trees in Urban North America

Cities throughout North America, and the world in general, are growing increasingly congested and polluted. This has prompted the need for viable strategies to mitigate the negative externalities of urbanization. The creation of urban forests, and urban green space writ large, has emerged as a leading candidate in this connection. This is especially true when urban forests and greens spaces are taken to connote all public and private open spaces in built areas. By definition, these areas are typically covered by green vegetation and are usually opened to the general public. They usually vary in size, location and intensity of use. However, a carefully planned green space, especially a public park, is one that is located at the heart of neighborhoods. This is necessary to ensure accessibility to all; more especially, it is necessary to avoid the problem of spatial injustice often perpetrated against socio-economic, racial/ethnic and other minorities.

Urban forests assume a unique meaning in North America especially because of the region's relatively recent transition from a mainly rural to a predominantly urban society. In just a little more than two centuries (1790s – 2000s), the level of urbanization in the United States, for instance, increased from 5% to 80% (US Census Bureau 2010). Canada, which has also registered high levels of urbanization, is currently 78% urbanized. The country, like the entire region of North America, has lost a significant part of its forest over the years. In fact, as recently as a century ago, the region was still predominantly rural. Presently, only one in five persons in the region is considered rural. Thus, an overwhelming majority of North Americans is based in urban areas. Although far well-off than people in other regions, most North Americans can neither afford recreational cottages nor have the luxury of experiencing the ecological, economic, and social benefits as well as the peace, tranquility, serenity and other advantages that trees and forests have to offer. Consequently, urban forests offer the vast majority of North Americans the only opportunity to reconnect with nature in general and the wilderness in particular. Thus, for most North Americans, the only forest experience they are likely to ever have is their interaction with woodlot remnants, community or public parks, highway and street trees. To fully understand the importance and role of urban trees and forests in North America, it is necessary to appreciate the rich history behind efforts to plant or safeguard trees and forests in built space in the region.

The history of these efforts predates the arrival of Christopher Columbus in the Americas in 1492. However, it is important to make the following notation: it is in North America, particularly in the United States that the concept of urban forestry initially took root. This, according to Gerhold (2007), predated the coinage of the term urban forestry and its eventual transmutation to a movement. More importantly, it is in North America that some of the earliest and best-known efforts to institutionalize the planting of trees and/or preservation of wild forest remnants in urban areas occurred. In this regard, Pennsylvania, a state whose name literally means "Penn's woods" (Igoe 2013) occupies a front row seat. Trees have always been of centrality since the genesis of the state. When William Penn was charged with the responsibility of crafting a blueprint for the City of Philadelphia in 1683, he made trees a central component. In this regard, he reserved land for five public squares in different parts of the city. Penn was influential in realizing environmentally-friendly and pro-tree policies in the city. One such policy required houses to be surrounded by gardens, fruit trees or orchards, and green fields. Penn's interest in trees was most manifested by the toponymic nomenclature that he adopted for purposes of christening the city's streets. Thus, streets were named after trees such as Cedar, Pine, Spruce, Walnut Chestnut, Mulberry (now, Arch), and Sassafras (now, Race).

A little more than a century, however, was to elapse before the residents of Philadelphia came around to making trees their business as opposed to the business of urban planners alone. In this respect, the actions that these residents took in 1792 are worth noting. These residents are said to have appealed to the municipal government to plant trees for their healthful effects (Gerhold 2007). During the years that followed, the City of Philadelphia became more actively involved in urban tree planting. However, it was not until 1896 that it hired its first professional arborist under the official title of Chief Forester. This was quite possibly the first time that urban forestry was institutionalized and afforded a visible place on a municipal government's organigram and list of priorities. Other municipalities in states such as Georgia, California, Massachusetts, New York, and Michigan soon followed Philadelphia's lead by establishing urban trees and/or urban forestry units of their own. Currently, such units constitute a ubiquitous element in municipal governments throughout North America. Commensurate with this has been the practice of urban tree-planting and forestry—activities that are carried out under the auspices of professionals from different disciplines, including but not limited to arboriculture, landscape architecture, and forestry.

Over the years, national, state and local governments have refined policies on trees. Thus, as Profous and Loeb (1990) have observed, many governments currently have laws and regulations that control the removal and conservation of trees in built space. The laws vary significantly from one polity to another. Some have argued that the variations are dictated by geographic, political, cultural and fiscal factors (Profous and Loeb 1990).

A common thread running through tree laws in North America is the fact that they all trace their roots to common law as derived from British Common Law. However, over the years, and especially in North it has been subsumed under certain statutes, and municipal by-laws. In general, the tree law governs all trees including privately, communally, and government-owned trees. Most jurisdictions in the region prohibit the cutting or pruning of designated trees without authorization. For example, the City of Victoria has a Tree Preservation By-law. Commensurate with this, the city has tree-preservation officers. Authorities are very strict with the enforcement of tree laws to the point of criminalizing any action that harms trees.

In Canada, the accentuation of urban tree planting traces its roots to the early 1960s. This is when the country's American elm *(Ulmus Americana L.)* a popular tree in North American built space was devastated by Dutch elm disease. Efforts to plant trees in Canadian built space date back to the turn of the twentieth century. This period marks the point at which municipalities throughout the country began taking steps to protect and plant trees to replace those lost to urbanization and related developments. As part of efforts in this regard, municipal departments and agencies were created throughout the country to manage vast tracks of public urban forests. The preferred agricultural strategy was monoculture, and institutionally, the emphasis was on the arboricultural industry. Some of the major actions in this regard included the creation of municipal tree advisory committees, the employment of professional foresters in most cities throughout the country. In addition, Canadian authorities embarked on preparing and enacting laws designed to govern the treatment of trees. One testament to the meticulous nature of these laws is illustrated by the manner in which they treat fruit trees in built space. In this regard, the law holds that a fruit tree owner had the right to enter a neighbour's property for the sole purpose of retrieving a fallen fruit without the risk of being charged for trespassing. This is, as long as the fruit-owner causes no damage to the neighbour's property. Commensurate with this, authorities established agencies to implement the laws and manage urban forests. The *Commission de la Capital Nationale* in Quebec City owes its origins to these early efforts. Also, counting among the very early institutions dedicated to urban forestry are the National Capital Commission based in Ottawa, the Wascana Centre Authority in Regina, Saskatchewan.

In the United States, municipal authorities have also been busy enacting tree laws. One distinguishing mark of these laws is that they differ by state, and sometimes, from one municipality to another. The case of laws governing the removal of trees on private property is illustrative (Kass 2018). Some municipalities require a permit to remove a tree on private property while others do not. The City of Birmingham, Alabama, Juneau, Alaska, exemplify municipalities in this second group. The cities of Sacramento, California; Tallahassee, Florida; and Miami, Florida are examples of municipalities in the first group. Some municipalities require permits only when removal pertains to certain species of trees. For instance, the City of Sacramento, California requires a permit to remove trees designated as heritage and set aside for preservation. An example of this is the oak tree, which is protected by California State Law. In the case of the City of Tallahassee, Florida, the tree law requires a permit to cut any healthy tree with a diameter of 36 inches or more on private property. The law in Miami, Florida requires a permit before removing more than 25% of the trees on any private property.

7.4.2 The Avowed Role of Trees in Built Space in North America

The practice of attributing monetary value to trees dates back to antiquity. In more recent times, the availability of trees on a parcel of land has been known to significantly increase the parcel's value. In ancient Greece, Plato had established a link between trees and soil and water conservation in watersheds. In ancient Rome and Greece, certain benefits of trees such as their ability to serve as windbreakers, pollution cleansers, and ground water replenishing agents have been recorded.

In North America, trees are recognized for having several merits in built space, particularly urban areas (Nowak et al. 2010). This explains the popularity of trees and other features of green infrastructure in the region's built space. These merits repose on ecological, social, economic, historical, cultural and institutional foundations.

Publicly and privately-owned and managed trees and woodlots in urban areas—that is, urban forests—have always constituted an integral part of the ecosystem of built space in North America. Urban forests typically include "individual trees along streets and in backyards as well as stands of remnant forest" (Nowak et al. 2010: 3). In 2010, it was estimated that US urban forests contain as many as 318 billion trees with an estimated asset value of 2.4 trillion (Nowak et al. 2010: 5). Yet, it is erroneous to compute the worth of these forests in monetary terms alone. This is because urban forests provide many other benefits to people and communities throughout North America. Proponents have advanced the following as prominent among these benefits in the specific case of North America (Nowak et al. 2010): moderation of local climate and energy use; improvement of air quality; mitigation of the negative impacts of climate change; enhancement of water flow; and noise abatement; amelioration of wildlife biodiversity; increase in real estate values; enhancement of individual health and wellbeing; improvement of community wellbeing.

Trees have been known to positively affect temperatures. In particular, "trees influence thermal comfort, energy use, and air quality by providing shade, transpiring moisture, and reducing wind speeds" (Nowak et al. 2010: 6). In the United States, for example, 100 million trees saved two billion dollars in the 1980s (Akbari et al. 1992). An established role of trees is to improve air quality, especially by lowering air temperatures and mitigating air pollution. Urban trees are said to remove some 784,000 tons of air pollutants annually with a monetary value of $3.8 billion (Nowak et al. 2010: 6). Trees in urban areas help fight against the effect of climate change. In this regard, trees store carbon within their tissues and reduce carbon emissions from power plants through lowered building energy use. Urban trees in North America are well-known for their ability to store a lot of carbon. According to one source, trees in North America reduced a total of 770 million tons of carbon valued at $14.3 billion in 2009 (Nowak et al. 2010: 6). The role of trees in increasing water flow and improving water quality is also noteworthy. Trees and natural vegetation in general help to retain underground water sources. In addition, trees help to hold soil in place, which in turn, removes harmful substances from water washed off roads and other surfaces. This nullifies the need for costly storm water recycling systems. Trees are known to serve as effective noise barriers. This is especially true for trees that are properly planted and spaced out; spaces of about 100 feet are ideal (Nowak et al. 2010: 6). Also worth noting is the ability of trees to promote biodiversity by providing a sound habitat for wildlife. Furthermore, trees provide an environment for preservation of endangered wildlife and plants.

Trees are well-known for their ability to remediate soils. This can be very important in places where soil quality has been degraded or is at risk of being degraded. Examples of such places include landfills and arid regions. Trees in urban areas have also been known to serve many purposes. However, one of these purposes, namely the increase of property values appears to be little-known. Yet, there is overwhelming evidence attesting to this highly positive attribute of urban trees. To be sure, the positive externalities do not accrue to real estate property alone. As one empirical study on Seattle revealed, the prices of assorted goods were systematically higher in areas with trees than in those without (Nowak et al. 2010: 7).

Urban trees have qualitative and quantitative values for the health and well-being of urban residents. Qualitatively, trees are aesthetically pleasing. Quantitatively, trees can enhance their exposure to intensive solar radiation. Urban trees of certain species have been known to have a high economic value. In addition, trees can provide a justification for neighbours to get together thereby promoting a sense of community amongst residents of urban communities.

7.5 Urban and Peri-Urban Agriculture in North America

7.5.1 Urban Planning and Urban Agriculture

Despite its propensity for insisting on spatial order through instruments such as zoning that segregate land use activities, urban planning has a history of encouraging agriculture in built space. A prominent fixture in this history is the garden city. As a concept connoting low density urban development, the garden city has often been misused especially by property developers to promote gated communities (Vernet and Coste 2017). However, the original concept as advanced by Ebenezer Howard (1898) was informed by the fervent belief that to be sustainable, a human community must be self-contained and self-sufficient with respect to food, recreation and transportation. Thus, the notion of "garden city as an autonomous human settlement" (Vernet and Coste 2017: 50). In its classic form as proposed by Howard, the garden city possesses four main features as follows: a greenbelt surrounding the community, proportionate areas of residences, industry, and agriculture. Thus, Howard had the foresight of appreciating the need to promote a harmonious co-existence of built space, including buildings, civil infrastructure and industries, and nature, mainly comprising farms, forests, and bodies of water such as rivers, streams and lakes.

Soon after it was launched in 1898, Ebenezer Howard's garden city model became popular among planning and municipal authorities throughout the world. In fact, many countries in the world borrowed and implemented the model in their urban planning endeavors. As a region, North America ranks as the leading importer of the model. More than a handful of cities in the region were planned as garden communities. Examples in the United States include, Mariemont, Ohio, one of the first garden cities in the United States; Greenbelt, Maryland, Greenhills, Ohio; and Greendale, Wisconsin. Others include Jackson Heights, Queens, New York; Reston, Virginia; Sunnyside Gardens Historic District, Queens, New York; Epcot, Bay Lake, Florida; Fairview, Camden, New Jersey; Forest Hills, Boston; Chatham Village, Pittsburg, Pennsylvania; and Radburn, New Jersey. In Canada, the best-known garden city communities include the following: *Citè-jardin du Tricentenaire* (i.e., the Tricentennial Garden City), Montreal, Quebec; Gardenvale neighbourhood, Quebec; Town of Mount-Royal, Montreal, Quebec; and Kapuskasing, Ontario. While some of these cities may only be garden cities by name today, the time and effort that went into their planning attest to a willingness to adopt Howard's principle of treating nature as part of built space.

This principle is fast regaining its once lost allure as there is a growing recognition of the importance of natural resources in built space, including villages, towns and cities. In fact, the need to promote this type of urban development constitutes the impetus for advocating urban and peri-urban agriculture (UPA) throughout the world today. The chief advocate of UPA, the Food and Agriculture Organization (FAO), defines UPA as "the growing of plants and the raising of animals within and around cities" (FAO 2019: para. 1). The importance of UPA cannot be overstated; it "provides fresh food, generates employment, recycles urban wastes, creates greenbelts, and strengthens cities' resilience to climate change." The focus and critical aim of UPA as currently promoted by the FAO differ by setting. In developing countries, the focus is typically on horticulture, livestock, fisheries, and agroforestry; and the aim is to combat hunger, produce fuelwood and generate employment. In developed countries such as the United States and Canada, the focus is on organically-grown food crops, including grains, vegetables as well as small animals such as rabbits, goats, sheep and non-food products such as aromatic and medicinal herbs, ornamental plants and other tree products. The main aim is to make healthy foods more affordable and accessible. If nothing else, this suggests that the food security problems of developing and developed countries are significantly different. In the former, the problem is often one of food scarcity while in the latter, it is hardly ever the lack, or scarcity of food. Rather, it is often the lack of healthy food. Amirtahmsebi (2012: para 2) paints a vivid picture of this problem with respect to the United States; she decries the fact that while 'junk food' such as hamburger is available for as a little as a dollar, thrice as much is required to purchase a bag of baby carrots. Paradoxically, therefore, food insecurity manifested by cheap and nutritionally inadequate food constitutes a major problem in developed countries such as the United States and Canada. A non-negligible segment of the population in these countries experiences food insecurity. In Canada alone, as many as 4 million persons suffer from food insecurity and related problems such as failing to take their prescribed medications regularly (Proof 2019).

Apart from helping to directly address the problem of food insecurity, urban agriculture plays three other important roles in the politico-economic context of North America. The first of these has to do with overcoming the grip of the industrial food system on the region's economy. The industrial food system is designed not to grow food to reduce food insecurity and promote environmental sustainability. Rather, its main purpose is to maximize profits for large food and cognate corporations. The second additional role of urban agriculture is psychological in nature, and has to do with re-introducing nature in people's lives. This is especially important on account of the fact that urbanization has increasingly elongated the physical distance between the natural and built environments in the region. The third is related to environmental stewardship and the need to reinforce the resilience of cities in the region to climate change and its multifaceted effects. Urban agriculture encourages recycling in many ways. For instance, domestic waste water can be used to irrigate urban gardens while grass from lawns can be used as compost manure in gardens. To gain a better understanding of the nature of urban agriculture in North America, it helps to examine what obtains in each of the two countries, the U.S. and Canada, comprising the region.

7.5.2 *Urban Agriculture in the United States*

Interest in urban agriculture has been growing in the United States during the last few decades. Evidence of this interest is found in the number of community farms that have been created throughout the country in the recent past. The Whitelock Community Farm discussed in Box 7.1 constitutes an example of such farms. Yet, knowledge on this important activity remains scant. In particular, not much is known with respect to what urban agriculture entails. Who is involved in it? Where it is typically done? What is the nature and magnitude of the challenges it faces? My aim in the following passages is to attempt to address these questions.

Borrowing from the FAO definition afore-quoted, the United States Department of Agriculture (USDA), characterizes urban agriculture as the practice of growing crops and raising animals within and around cities. Thus, an urban farm is any place, such as a piece of land, an abandoned parking lot, a balcony or backyard on which crops are grown or animals are raised. The USDA proffers a narrower definition of what constitutes an urban farm. It characterizes this as "any place from which $1,000 or more of agricultural products were produced and sold or normally would have been sold during the census year" (ATTRA Online, para. 2). This definition gives the erroneous impression that, to be an urban farm, it must generate surplus that can be sold. Yet, as discussed later, many Americans practicing urban agriculture do so for recreation or non-commercial purposes.

To be sure, those with a stake in urban agriculture are not only individuals, households or families. Individuals who are on average, younger than rural farmers, continue to play a vital role in the urban farming sector. However, it is worth noting that the stakeholder pool of urban agriculture extends beyond individuals; it com-

Box 7.1 Whitelock Community Farm: An Urban Farm

Whitelock Community Farm in the Reservoir Neighborhood, Baltimore, Maryland, USA exemplifies an urban farm by a community for its members. The organization's main objective is to serve as a dependable source of fresh food for residents of Reservoir and proximate neighborhoods. It helps to reduce unemployment by providing temporary part-time, and occasionally, full-time employment opportunities to community members. Furthermore, it seeks to promote urban agricultural activities such as gardening, composting and horticulture. To support its avowed mission, it organizes farm-based learning programs, including but not limited to classes and workshops for children and adults. The organization holds membership in the Baltimore City Farm Alliance. This is a network of agro-producers striving to increase the sustainability of urban farming with a view to increasing the availability of urban-grown foods.

Source: Rosenberger (2016).

prises a wide range of entities, including federal government agencies such as the United States Department of Agriculture (USDA), municipal authorities, city governments, food policy councils, religious bodies, and local community groups. Cities such as Baltimore, New York, and Chicago have been active in promoting urban agriculture to the point of incorporating suggestions from citizens to expand their urban agriculture sector (ATTRA 2013). Community groups are playing an increasingly important role in the urban agriculture arena. Box 7.1 discusses the gigantic role that one such a community group has played producing fresh food, and generating employment for members of the wider community to which it belongs.

A closer look at who is actively involved in the urban agriculture domain as reported by a recent national survey of urban farmers is very informative (see ATTRA 2013). As stated above, urban farmers in the United States is on average younger than rural farmers; they are typically between the age of 21 and 78; and; are mostly (53%) female. The overwhelming majority (91%) are Caucasian, while 5% are Black, 4% Hispanic, 3% Native American, and 1% Asian. These statistics must be interpreted with caution. For one thing, the categories are not mutually exclusive. The survey questionnaire items allowed for the possibility of respondents checking more than one category. For another thing, the items made no distinction between participants who conducted urban agriculture as individuals or part of a group. Such a distinction would have been illuminating with respect to the extent and role of community participation in urban agriculture in the United States. Finally, because of the vagueness of what constitutes an urban area in common parlance, it is difficult to disaggregate the data elicited by the questionnaire into 'rural' and 'urban.' However, the survey sheds intense light on the land ownership status of urban farmers in America. More than half of the farmers own their urban farm plots. Of these, 64% reported that some of the land was under public control—in other words, public land, about 27% (27.5%) possessed a long-term lease, while a significant proportion (37%) has farms at multiple sites. The effect of land ownership status on the success of urban farming cannot be overstated. Land tenure status affects the duration of farming activities, which in turn impacts the farmer's investment decisions. Such decisions are consequential for those practicing urban farming as an avocation rather than a hobby. Here, it is necessary to note, as the survey revealed, that most American urban farmers fall under the latter category.

One question that often surfaces in the discourse on urban farming relates to the scarcity of space. How does one come about space to farm in urban areas, which by definition are densely populated. Lessons on the farming strategies employed by American urban farmers as reported by the National Center for Appropriate Technology (NCAT) (see ATTRA 2013) can go a good way in uncovering answers to this question. Most (64%) of the urban farmers surveyed employed raised beds, while 41% employed greenhouses, and 37% employed soil containers. Other strategies or techniques employed include, 'high tunnels' (29%), vertical farms (17.8%), aquaponics (7.6%), hydroponics (5.4%), and rooftop farming (3%). Another noteworthy question relates to what region in the United States one is likely to find an urban farm. The study reported that the Northeast, with 32% of urban farming activities, leads the country; followed by the South (26%), then, the West (22%), and finally, the Midwest (18.8%). These statistics are very telling, and easy to explain. The Northeast is the most densely populated and most urbanized region of the country. It is also the most cosmopolitan, with members of its population drawn not only from different parts of the country but different parts of the world. These factors produce two likely results that bode well for the practice of urban farming. First, with their elongated distance from rural areas, residents of the Northeast are likely to practice urban farming as a means of literally drawing nature closer to themselves. Second, and closely linked to the first, urban farming offers residents of the Northeast, especially those from farming parts of the United States or the world at large, an opportunity to continue a vocation they are used to. For immigrant in the Northeast, therefore, urban farming affords them the opportunity to create in their host country, a home-away-from-home by continuing the farming lifestyle they grew used to in their native countries. The high incidence of urban agriculture in the South can also be explained by identical factors. The south, especially states such as Florida and Texas are popular destinations for immigrants, particularly those from Latin America. A more likely explanation is the fact that what passes for an urban farm in the region today is likely a previously rural farm, which has recently been circumscribed by urban development.

Although oft-ignored, a motivating factor behind urban farming in the United States is economics. A study by the National Centre for Appropriate Technology (NCAT) revealed that mid-sized urban farms reported sales of over $350,000 in 2012 (ATTRA 2013). It also reported that urban farmers market their surpluses through many outlets, including Farmers' Markets, regional and local food hubs; also, they sold their products directly to wholesale outlets, retail establishments, cooperatives and restaurants. However, urban farmers, including those farming for pleasure and commercial reasons, continue to face many challenges. Prominent among these are gaps in knowledge of factors that determine the effectiveness of urban farming, institutional barriers, including urban planning laws and regulations, especially zoning and building codes. These laws and codes are especially impactful on farms within city limits. In addition, urban farmers face many hurdles with respect to procuring credit and capital. Furthermore, urban farmers must wrestle with other problems such as the following: complicated land tenure formalization procedures, lack of infrastructure for marketing, difficulties in processing food, limited access to water, problems related to environmental contamination and lack of technical assistance for urban farmers.

7.5.3 Urban Agriculture in Canada

To appreciate the role of urban agriculture in Canada, one must first understand a few basic facts about the country. The country's population, which stood at 37 million in 2018, is distributed mainly among 147 metropolitan areas. Some have characterized the population distribution as congregated in groups ranging from 5,600,000 in Toronto to 10,500 in Bay Roberts, Newfoundland (Levenston 2016).

The urban agriculture domain in Canada is particularly noteworthy because of the support it receives from the state. This support runs deep in the country's history. In fact, this practice goes as far back as to the time the first settlers arrived French Canada. An important fact of these early settlers is that they were encouraged by the government to grow potatoes as a means of subsidizing their diets. More recent efforts to promote urban agriculture in the country occurred in the 1940s, and precisely during World War II. This is when the government urged urban residents throughout the country to grow food at their homes in order to help with the War effort. The government's initiatives in this connection included the production of flyers, posters, and other documents beseeching citizens to indulge in urban agriculture. One such poster carried the following message (Levenston 2016: 1)

> Grow these 'fighting foods" at home. It's a job we all can do; The fresher food will do you good; And you'll help your country too.

The poster bears the picture of potatoes, carrots, onions, and tomatoes. These, the poster was designed to convey, are foods that can be easily grown at home. Home-grown food is important "for the health of all Canadians; to people with garden-fresh food; to improve the environment; and to lift our spirits" (Levenston 2016: 2). The information transmitted through the poster and other public pronouncements, attest to the Canadian government's commitment to urban agriculture. At the peak of the war in 1944, as many as 20,200 farms, which were fondly known as 'Victory Gardens,' were in operation throughout Canada (Levenston 2016: 2).

The Canadian government has also been active in the urban agriculture domain through public educational and research institutions. One institution that stands out in this regard is Canada's International Development Centre (IDRC). This entity has been a major actor in the urban agriculture arena for several years. Its role has mainly been in the areas concerned with agricultural research. In this regard, one of the goals has been to promote knowledge of urban agriculture. As part of efforts in this connection, it undertook an ambitious program known as 'Cities Feeding People' between 1996 and 2005. In addition to government agencies, non-governmental and/or non-profit organizations have also been active in the urban agriculture domain in Canada. One example is the non-governmental organization (NGO) known as the City Farmer Society in Vancouver. Its main role has been to research and promote urban agriculture, activities it has been involved in urban agriculture for more than four decades. The Canadian government's support for urban agriculture is only one indication of urban agriculture's popularity in Canada.

Another indicator can be found in the many sites that urban agriculture occupies throughout the country. The following examples are illustrative (Levenston 2016). The first is the Parliamentary buildings in Quebec City. This is comprised of five thematic sections and 130 different plant variety on 'smart plots.' The second is the Food Bank, which is Saskatoon's flagship project also known as the Garden Patch. This is a community-driven urban agriculture project; it sits on a vacant city-owned lot that has been converted to a vibrant green garden each year. It is on record for generating 75,000 pounds (about 34,019 kg) of farm produce between 2012 and 2016. The third is the University Rooftop Project located at Ryerson University in

Toronto. Also known as the Edible Rooftop Garden, this facility was developed by converting an existing green roof into an urban farm by adding compost, sheet mulch and various seedlings from the Ryerson University greenhouse. The fourth is the Front Yard Garden, Drummondville, Quebec. This is a farm owned and operated by a married couple. They grow mainly vegetables. However, this case exemplifies one of the main barriers to urban farming in North America. In this case, the Quebec municipal authorities find the farm in contravention of urban development by-laws and has since issued them a notice to significantly reduce the farm's size or shut it down completely. The authorities have promised to fine the couple $300 per day should they fail to comply. The fifth is the Community Greenhouse, INNVIK, which has been characterized as "the most noteworthy greenhouse in North America. It was opened in 1999. Today, it contains 174 full-size plots. Each plot measures 8 feet by four feet, and rent for $50. Members grow an assortment of vegetables, including, but not limited to spinach, chard, and lettuce as well as tomatoes, carrots, peas, herbs, strawberries, rhubarb, zucchini, and squash. The sixth is Backyard Garden in Vancouver. Most homes in Vancouver have always maintained gardens in their backyard. Throughout history, new immigrants to the city have brought along with them farming skills. The seventh is the Commercial Rooftop Farm in Montreal. It was constructed by a private corporation in 2011. It stands out as the world's first commercial rooftop greenhouse. It measures 32,000 square feet, and sits atop a majestic building in Ahuntsic, Montreal. The eighth is the Refugee Garden in Halifax. This is an urban garden located on land donated by the Mosaic Ministries, an independent church in Halifax. The farmers are mainly from Asia, particularly Bhutan. The ninth is Allotment Garden in Burnaby. This urban garden was established by the Government of British Columbia in the 1970s. this farm grows vegetables such as berries, lettuce, and spinach; in addition, it grows flowers. This 5.9-hectare piece of land contains 372 garden plots, each of which measures 1000 square feet—that is 50 feet by 20 feet. The tenth is the Hotel Ford Garden. This comprises a group of ten Fairmont Hotels and Resorts, which own and operate their own urban gardens. They grow fresh herbs and vegetables that they prepare and serve their clients in Banff, Jasper, Whistler, Vancouver Toronto, Quebec and Montreal.

7.5.4 Challenges to Urban Gardening in Canada

The urban gardening sector in Canada faces a number of challenges, some of which are ubiquitous and others which may be peculiar. Some of the challenges that face urban agriculture throughout North America include the fact that a significant proportion of urban dwellers live in multi-family housing units. This means among other things that there is a scarcity of space for agricultural purposes. Consequently, some degree of creativity is required. In this regard, spaces such as balconies, rooftops, vacant public land, and spaces such as school yards that are unused during some periods of the year have been proposed as fertile ground for urban agriculture.

Yet, space is not the only problem that urban farmers face in Canada. Another critical problem relates to municipal laws and regulations, especially those having to do with activities permitted within the city limit. Consider the case of poultry farming. This activity is permitted in some cities but not others. The Canadian cities that permit poultry farming include Vancouver, Victoria, Kelowna, Surrey, Montreal Gatineau, Moncton, and Whitehorse. A few large cities forbid this activity outright; they include, Toronto, Ottawa, Calgary, Halifax, Winnipeg, Regina and Saskatoon. A number of smaller cities also prohibit poultry within the city limit. These include Grand Prairie, Airdrie, Peace River, and Fort Saskatchewan in Alberta; and Guelph, Waterloo, Brampton, Quinte West and Niagara Falls, Ontario.

7.5.5 Addressing Challenges and Problems in Urban Agriculture

The challenges identified above have not escaped the attention of the Canadian government, which has embarked on initiatives to accordingly address them. The following ten initiatives emerge as the most prominent among those that have been recommended or taken by the Canadian state (Levenston 2016). The first is the recommendation for the state to create a Federal Office of Urban Agriculture. The main responsibility of this office will be to coordinate national efforts in the urban agriculture domain. In addition, it will be responsible for collecting and disseminating data on urban farming, including the number of people involved, amount of land available, the urban farming sector's productivity.

The second is for the Federal Government to convene a roundtable of Canadian experts. These experts will provide the government with knowledge and information most necessary to promote urban farming. The third recommendation is for the government to establish a model food garden on Parliamentary Hill. This will seek to accentuate the government's commitment to urban gardens. Fourth, expert opinion suggests that the government would do well to establish demonstration gardens. These gardens will serve as education and information hubs for residents of large municipalities. An important function of this entity will be to teach residents urban farming techniques. Fifth, it is recommended that the Canadian government should convene a national conference on urban agriculture. The Federal Office of urban agriculture will invite Canadians to a national networking event. Urban farmers practicing their craft at home, school yard, rooftop, and community gardens as well as commercial urban farmers will learn from each other. The sixth recommendation is for the government to create a national urban agriculture website. This is necessary to serve as a conduit for disseminating crucial information for urban farming. Seventh, the government has been urged to actively coordinate urban agriculture policies. The Federal Office will be responsible for policies for city food production. Examples of the type of specialized urban farming requiring special attention include beekeeping and poultry farming. The eighth recommendation is for the government to provide funds and other resources to help defray the cost of urban farming projects. Urban farming requires much by way of resource input. These resources, such as phone apps, and websites require resources. The ninth recommendation is for the government to actively facilitate linkages between urban and rural farming. The viability of the link between urban and rural farming, and by extension, urban and rural farmers, is contingent upon a recognition of the fact that such a connection can facilitate the exchange of information between the two entities as well as international parties. The final recommendation is for the Canadian government to facilitate international urban agriculture connections. This is necessary to facilitate information exchange between Canada and other countries especially with respect to urban agriculture strategies.

7.6 Urban Green Energy

7.6.1 Current Renewable Energy Situation in North America

North America is among the world's leading consumers of energy. It is therefore, no wonder that the region counts as one of the world's leading contributors to problems such as air pollution, environmental degradation and climate change. These problems are associated with energy use. Paradoxically, it is the region with the economic, scientific and technological capacity required to eradicate these problems. However, due mainly to political reasons, these resources have never been fully committed to efforts designed to eradicate the problems. For example, the region has been slow in making the transition to renewable energy sources. Consequently, it continues to rely heavily on fossil-fuels. In this regard, coal-fired plants remain a dominant source of power in the region. Yet, there are signs that coal is losing its once indisputable allure. According to the US Energy Information Administration (EIA), The US recorded a drop of 3.6% or 1020.5 million short tons (MMst) in coal production between 2017 and 2018 (US EIA 2019, para 3). At the same time, coal consumption decreased by 4%. Despite the slight decline in its production and consumption, coal continues to have far-reaching implications for built space in North America. To appreciate these implications, consider the fact that most (92.6%) of the coal consumed in the region goes to the electric power sector. Reversing this situation requires active efforts to improve access to renewable energy in the region.

Currently, renewable energy accounts for 11% of the total energy consumption in the United States, for instance; most of this comes from hydro sources. It is also the source of about 46% of the total renewable electricity generated in the country. Hydro sources currently constitute the leading source of renewable energy in the entire North American region. Here, it is important to note that the two countries in the region rank highly among top producers of hydro-electricity in the world. In this regard, the United States ranks fourth, while Canada ranks second—after China but ahead of Brazil. The region boasts some of the world's richest renewable energy sources. In particular, the region has the best potential to maximize the utility of renewable energy. Prominent among the renewable energy sources in the region are wind, solar, geothermal, hydro- and biomass. These sources bode well for efforts to shift from conventional to renewable energy in the region. Other crucial factors include the region's culture of innovation and entrepreneurial spirit, vast financial opportunities, and a highly-skilled labour force.

Renewable energy has speedily risen to become a significant source of energy in North America. Currently, the region relies on renewable energy for large-scale power generation. This is especially in the form of hydropower. The case of Canada, where hydropower accounts for 63% of electricity generation, some dams in the country are as old as a century (IRENA 2019, para. 2). The U.S. is also dependent on hydropower for a significant portion of its electric energy. In fact, the U.S. is the home of iconic dams such as the Hoover Dam on the Colorado River. It is necessary to note that this dam is one of the pioneering hydropower sources in the world. In recent times, other renewable energy sources have been making their mark in the renewable energy domain. Leading this trend is solar energy, which currently employs more than 260,000 people and create jobs at a rate significantly higher than the country's average.

The 'green economy' in general has been an enormous boon to the North American economy in general. For instance, despite political rhetoric from conservative quarters in the United States that paint a negative picture of environmentalism, the country's 'green economy' is worth $1.3 trillion. The contribution of the renewable energy sector and the 'green economy' writ large has assumed many forms. For instance, there has been a rise in clean energy-related jobs. In this regard, as many as 9.5 million Americans, that is, about 4% of the U.S. workforce is employed in a 'green economy' activity. This sector accounts for about 7% of the country's GDP (Bloomberg 2019). The U.S. has historically depended on coal for energy; in fact, coal has been the dominant source of energy for the country's industrial and domestic sectors. However, as a recent report by the Cable News Network (CNN) stated, renewable energy recently surpassed coal as the country's leading energy source. In particular, the renewable energy sector registered a slightly higher installed capacity than coal. This means that U.S. power plants are, for the first time in history, able to produce more energy from clean sources than coal. One factor accounting for this phenomenon is the rapidly falling cost of energy from renewable sources such as the sun and wind. Another is the fact that Americans are drastically curtailing their voracious appetite for coal, the consumption of which peaked in 2008. It has been on a decline since then. Recently, the consumption of coal decreased by 39%, which is the lowest level in four decades. Prior to this, natural gas had surpassed coal as the preferred energy source in the country in 2016. This was occasioned by the so-called shale boom, which resulted in an abundance of cheap natural gas.

7.6.2 *Energy Policy Initiatives in North America*

Despite its heavy reliance on non-renewable energy sources, the North America region is a world leader in efforts to combat threatening energy-related environmental problems such as global climate change. For example, significant grounds have been gained in efforts to eliminate the production and use of ozone-depleting substances. Some remarkable success has also been registered in the region with respect to cutting carbon dioxide, sulfur dioxide and other emissions with air polluting implications. Examples of specific policy initiatives include recent pieces of legislation such as Carbon Tax and Cap-and-Trade. The Carbon Tax has been in effect in places such as Quebec and British Colombia since 2007. In practice, this tax requires Energy companies to pay 0.8 cents for each liter of petrol (gas) distributed in Quebec and 0.938 cents more for each liter of diesel. Essentially, this encourages the distribution of gas (a cleaner energy source) but discourages the distribution of diesel (a dirtier source). Other policies that have registered significant positive results have targeted transmission networks. Energy use has been significantly reduced in this area by designating transmission cost recovery and allocation, and managing the grid through independent operators, and developing smart grids.

7.7 Water Supply in Built Space

7.7.1 *The State of Urban Water Supply*

Despite its level of financial, human and natural resource endowment, North America faces a few nagging problems in its water domain. To appreciate these problems, it helps to understand the main sources of water in the region. The World Atlas (Online) has shed some light on this on these sources. It notes that North America's drinking water comes mainly from the following sources lakes, rivers, aquifers, aqueducts, and reservoirs. The United States alone boasts at least 100,000 lakes and more than 250,000 rivers as well as hundreds of reservoirs (World Atlas Online, para. 2). These latter are the sources of freshwater for many cities throughout the country. However, it is important to note that much work goes into treating the water before it is considered ready for consumption. Thanks to this, North America's drinking water is considered one of the best in the world. The region is well ahead of the United Nations (UN) in recognizing safe drinking water and sanitation as a basic human right. Authorities in the region had taken steps to guarantee access to water for all long before the UN General Assembly recognized such access as a basic human right on 28 July 2010. In the United States, for instance, the Clean Water Act, which guarantees access to safe water was enacted in 1972, while the Drinking Water Act with a similar goal was enacted 2 years later in 1974. To ensure that the water supplied to the population is always safe, it must meet both federal and municipal standards.

Ground and surface water constitute the most important source of water for drinking and domestic use and irrigation in the region. In the US, for instance, 60% of irrigation water comes from the ground. The implications of the heavy dependence on ground and surface water sources are far-reaching. For example, extended periods of droughts on the one hand, and heavy demand on the other almost invariably leads to overexploitation and eventually, irreversible damage such as aquifer depletion. The infamous California droughts affecting the High Plains Aquifer and California Central Valley Aquifer are illustrative. These problems affect significant parts of the U.S. population to the point of making scarcity the country's leading water problem—one that is exacerbated by climate change and resultant extreme weather events. Apart from the droughts affecting southern California, the impacts of climate change are already being felt in other places. For instance, there has been a noticeable drop in the volume of large bodies of water such as Lake Mead, Arizona and the Colorado River, Colorado. Climate change has also been identified as the cause of the decreasing water containment in the Colorado basins such as Lake Powell, as well as the lower course of the Colorado River around Baja in California. In fact, this latter portion of the river is already running dry. To appreciate the impact of this on built space, consider the fact that as many as 22 million people depend on Lake Mead to meet their water needs (Snyder 2019). These people, and other populations depending on sources of water that are dwindling are being asked to significantly curtail their water use. For instance, as Snyder (2019) has observed, homeowners who have, albeit illegally, drawn water from the Colorado River for decades have been issued orders to cease doing so or face severe fines. More noteworthy, people in such areas, particularly those along the arid Southwest are likely to face severe difficulties meeting their drinking and irrigation water needs within the foreseeable future.

Canada is experiencing similar phenomena; many rivers in the country are dropping in volume. Some are actually disappearing. The case of the Slims River in Yukon, British Columbia is illustrative (see Independent 2017). The recent work of Shugar et al. (2017) provide a logical explanation for this paradoxical phenomenon. The river's vanishing was due mainly to the melting of glaciers that used to feed it. The glaciers melted and retreated to the point of carving an alternative route to the sea. Such an event, it must be emphasized, is by no means commonplace. To be sure, this is the first time in 350 years that an entire river vanished in Canada. Yet, the Slims River is only one of many rivers under threat from climate change and commensurate problems. The Kaskawulsh Glacier, another important source of fresh water in the country, has retreated about a mile up its valley over the last 100 years (CTV 2009). In addition, Shugar et al. (2017) project that the glacier in western Cordiltera Mountain region which covers the Yukon, British Columbia, and parts of Alberta are on track to lose 74% of their volume by the end of the century. The impact of climate change is limited to small rivers; many of Canada's large rivers such as the St. Lawrence and the South Saskatchewan are also experiencing the effect of increasingly warming temperatures. These have resulted in slight increases in the temperature, and reductions in the volume, of their water. These changes have taken a toll not only on aquatic life but also on the ecosystem of the region writ large. More importantly for the purpose of this book, the changes have far-reaching implications for human populations and built space in general. In this regard, it must be noted that the changes are likely to prove disastrous for freshwater supplies and hydro-electricity generation.

7.7.2 Urban Water Supply Issues

The major water issues in North America can be summarized under the broad headings of scarcity, pollution, rising sea levels, warmer temperatures, affordability for the poorest members of the population, and an ageing workforce. Although a problem of significant magnitude, water scarcity is evenly distributed across the region. Thus, the problem is characterized by considerable variability across the region, depending on the difference in intensity of rainfall (Cromwell et al. 2007). Pollution also emerges as a significant problem in the region. Leading sources of polluting agents in water in the region are industrial wastes, processes and chemicals; sewer overflow, and pesticides and fertilizers employed in agriculture. As for rising sea levels, this is an element of the climate change problematic mentioned earlier. Oceans expand due to rising temperatures. This results in the melting of glaciers and ice sheets. This causes sea levels to rise. Warmer and potentially drier summer seasons. Other direct effects of increased temperatures include the following: increased evaporation in surface water sources; water treatment and distribution challenges (disinfection) by-product regrowth); increased water demand, which includes increased irrigation demand in longer growing seasons, increased urban demand with more heat waves and dry spells, increased drawdown of local groundwater resources to meet the above.

Policy Initiatives The strategies that have been employed to address the water problematic in Canada and the United States fall into two distinct categories, namely supply side and demand side. The supply side strategies that have been historically employed include, tapping into distant sources, especially rivers. However, in recent times, there have been increasing efforts to explore non-conventional sources such as seawater desalination in coastal areas.

An oft-ignored supply-side strategy that is growing increasingly popular, is water reuse. Reuse of reclaimed water is a ubiquitous strategy for responding to water scarcity problems in North America. This entails recycling and re-using water from residential facilities. Such water is typically used for irrigation purposes. Occasionally, it is used for car washing, industrial plant cooling, and the restoration of wetlands. A controversial water re-use strategy is known as Direct-Potable-Reuse (DPR). This involves households using chemically recycled water for domestic purposes, including drinking, cooking, cleaning and irrigation. The City of Wichita Falls, Texas was the first to adopt this controversial strategy in 2014. Re-use water has also been employed for the purpose of replenishing depleted groundwater sources. In practice, efforts in this connection usually include the digging of ponds or underwater receptacles. Characterized as inherently 'low-tech,' this strategy has been employed with great success by farmers (Snyder 2019).

Canadian and American authorities have also employed demand side strategies. Prominent among these have been efforts to improve water distribution and conservation systems. Some cities, such as Seattle in Washington, have actually succeeded in reducing per capita water use. Conservation programs can be comprehensive to include pricing policies, education, regulations, and rebates for water appliance. Cities such as Atlanta, Georgia and Las Vegas, Nevada, have also implemented water conservation programs. Las Vegas has intentionally focused on curtailing outdoor water demand. This latter accounts for 70% of residential water use in the city. Although seldom mentioned, demand-side strategies, especially those that have water conservation as a critical element, can only succeed when they recognize the crucial role of community participation (CP). Thus, as Snyder (2019) suggests, water conservation and volume promotion requires not only the input of government agencies but also that of water users or consumers. To appreciate the crucial place of individual consumers or households in demand-side strategies, imagine how much water can be saved by effectuating the following simple behavior modifications: refraining from the wasteful use of water, curtailing daily water demand, and using only the most effective plumbing and water-related appliance, refraining from using leaking faucets and toilet systems. The importance of small fixes such as these are well recognized in North America. This recognition is founded on knowledge of the fact that current water demand affects future availability.

7.8 Concluding Remarks

With only two countries, North America constitutes the least socio-politically diverse of United Nations Environmental Programme (UNEP) regions. However, it is geographically vast, and ranks as the most economically and industrially developed. These conspicuous features have far-reaching implications for nature in the region's built space, particularly its urban areas. This chapter has analyzed these features, identified and discussed some of the significant ways in which they affect nature in built space in the region. The focus is specifically on the major politico-economic, socio-cultural geo-ecological and historical factors facilitating or obfuscating efforts to promote nature in the built environment. The region's liberal democratic and economically advanced system, combined with its dominant Judeo-Christian culture, have proved to be a boon for efforts to promote nature in built space. Democracy as practiced in the region grants almost limitless autonomy to municipal governments. This accounts for the inter-municipal differential in policies affecting nature in built space in the region. Its vast pool of financial and human resources has facilitated execution of the task of creating and maintaining nature in the built environment. The most important factors shaping efforts to promote nature in built space in the region are of the ecological variant.

Problems originating from activities in the built space of North America have been known to cause environmental problems elsewhere. For instance, acid rain caused by factories, industrial plants in the region have occasionally wreaked havoc on the natural environment and built space in Latin America. Nuclear waste from power plants Nuclear, proven to be indestructible and indecomposable, and has consequences that transcend the US frontiers.

References

Akbari H, Davis S, Huang J, Winnet S (1992) Cooling our communities: a guidebook on tree-planting and light-colored surfaces. U.S. Environmental Protection Agency, Washington, DC

Amirtahmasebi R (2012) The North American urban agriculture experience. World Bank Blogs. Accessed 13 Oct 2019 via: https://blogs.worldbank.org/sustainablecities/the-north-american-urban-agriculture-experience

ATTRA (2013) Urban agriculture in the United States: baseline findings of a nationwide survey. Accessed online 14 Oct 2019 at: http://www.attra.ncat.org

Bloomberg E (2019) America's 'green economy' now worth $1.3 trillion, says study. Accessed 20 Oct 2019 via: bloomberg.com/news/articles/2019-10-15/america-s-green-economy-is-now-worth-1-4-trillion

Cromwell JE, Smith JB, Raucher RS (2007) Implications of climate change for urban water utilities – main report. Washington, DC: Association of Metropolitan Water Agencies (AMWA). Available online. Accessed 20 Oct 2019 via: https://www.amwa.net/galleries/climate-change/AMWA_Climate_Change_Paper_12.13.07.pdf

CTV (2009) Climate crisis putting Canada's Rivers at risk. CTV News. Available online. Accessed 20 Oct 2019 via: https://www.ctvnews.ca/climate-crisis-putting-canada-s-rivers-at-risk-1.444024

Durhaime.org (Online) Tree law. Accessed 8 Oct 2019 via: http://www.duhaime.org/LegalResources/RealEstateTenancy/LawArticle-671/Tree-Law.aspx

Ehrman B (2018) The triumph of Christianity: how a forbidden religion swept the world. Simon & Schuster, New York

Encyclopedia Britannica (2019) The impact of the motor vehicle. Available online. Accessed 29 Sept 2019 via: https://www.britannica.com/place/United-States/Impact-of-the-motor-vehicle

FAO (2019) Urban agriculture: FAO's role in urban agriculture. UN Food and Agricultural Organization (FAO). Accessed 15 Oct 2019 via: http://www.fao.org/urban-agriculture/en/

Gerhold HD (2007) Origins of urban forestry. In: Kuser JE (ed) Urban community forestry in the northeast. Springer, Dordrecht

Howard E (1898) Tomorrow: a peaceful path to real reform. Swan Sonnenschein & Co., London

Igoe LT (2013) Trees. In: The encyclopedia of Greater Philadelphia. Available online at: http://philadelphiaencyclopedia.org/archives/trees-2/

Independent (2017) Climate change causes huge Canadian River to vanish in four days. The Independent, Tuesday 1 18 April 2017. Available online. Accessed 20 Oct 2019 via: https://www.independent.co.uk/environment/climate-change-slims-river-river-vanish-canada-four-days-global-warming-little-ice-age-piracy-a7688421.html

IRENA (2019) North America. A web-based article by the International Renewable Energy Agency (IRENA). Accessed 5 Nov 2019 via: https://www.irena.org/northamerica

Kass BL (2018) Tree laws differ among states. Real Estate News. Accessed 19 Oct 2019 via: http://www.ygrouprealty.com/real-estate-news/Tree-Laws-Differ-Among-the-Sta-.htm

Levenston M (2016) Canada A national strategy for urban agriculture. Chapter 1: sunny days in the food garden

McHarg I (1971) Design with nature. Natural History Press, Garden City/New York

Mundi Index (Online) Canada economy profile 2018. Accessed 2 Oct 2019 via: https://www.indexmundi.com/canada/economy_profile.html
Nowak DJ, Stein SM, Randler PB, Greenfield EJ, Comas SJ, Carr MA, Alig RJ (2010) Sustaining America's urban trees and forests. Report for the U.S. Department of Agriculture, Forest Service. Northern Research Station. General technical report no. NRS-62, June 2010
Pew Research Center (2019) 6 Demographic trends shaping the U.S. and the World in 2019. Factank News in the numbers. Accessed 2 Oct 2019 via: pewresearch.org/fact-trends-shaping-the-un-s-and-the-world-in-2019/
Profous GV, Loeb RE (1990) The legal protection of urban trees: a comparative world survey. J Environ Law 2(2):179–193
Proof (2019). Publication: prescription medication nonadherence associated with food insecurity: a population-based cross-sectional study. PROOF, Food Insecurity Policy Research, University of Toronto. Accessed 13 Oct 2019 via: https://proof.utoronto.ca/publication-prescription/
Rodrigue J-P (2017) The Geography of transportation systems. Routledge, New York
Rosenberger A (2016) Twelve organizations promoting urban agriculture around the world. The Christian Science Monitor. Accessed 5 Oct 2019 via: https://www.csmonitor.com/Business/The-Bite/2016/1218/Twelve-organizations-promoting-urban-agriculture-around-the-world?cmpid=mkt:ggl:dsa-np&gclid=Cj0KCQjww7HsBRDkARIsAARsIT7U5cmkV3gVO3yQBJhBeHguVs-MPTYNeGgZOc5GJxViLDi1Ej-YON0aAvwdEALw_wcB
Shugar DH, Clague JJ, Best JL, Schoof C, Willis MJ, Copland L, Roe GH (2017) River piracy and drainage basin reorganization led by climate-driven glacier retreat. Nat Geosci 10:370–375
Snyder S (2019) Water scarcity—The US connection. Online publication, The Water Project. Accessed 5 Oct 2019 via: https://thewaterproject.org/waterscarcity/water_scarcity_in_us
Tarr JA (Online) The city and natural environment. Accessed 28 Sept 2019 via: gdrc.org/uem/doc-tarr-html#tarr
UNEP (2016) Global environmental outlook: regional assessment of North America. Nairobi. United Nations Environmental Programme (UNEP), Kenya. Also available online at: http://wedocs.unep.org/bitstream/handle/20.500.11822/7611/GEO_North_America_201611.pdf?sequence=1&isAllowed=y
US Census Bureau (2010) 2010 census of population and housing unit counts. CPH-2-5. US Census Bureau/Government Printing Office, Washington, DC
US EIA (2019) Annual coal report (Oct. 3, 2019). Accessed 10 Oct 2019 via: https://www.eia.gov/coal/annual/
Vernet N, Coste A (2017) Garden cities of the 21st century: a sustainable path to suburban reform. Urban Plan 2(4):45–60
White L (1967) The historical roots of our ecological crisis. Science 155:1203–1207

Chapter 8
Nature in Built Space in Western Europe

Abstract As analyzed in this book, Western Europe as a region coincides approximately with the European Union. It is the birthplace of the Industrial Revolution. It is therefore not surprising that it faces some of the most serious problems rooted in air pollution and environmental contamination writ large. These problems are present mainly in urban centres, which tend to be dominant given the region's status as the most urbanized in the world. As the most urbanized region, biodiversity loss emerges as one of the region's leading environmental problems. The region also faces some of the most stuborn challenges when it comes to developing and preserving nature in the built environment. While these problems are not new, their exact nature and origins remain largely unknown. This chapter seeks to shed light on the problems by analyzing them as a function of political, economic, technological, ecological, cultural and historical factors. In the process it shows how these factors have affected, or can affect, efforts to develop and manage nature in the region's built environment.

8.1 Introduction

This chapter focuses on Western Europe, which comprises the following twenty-three countries: Andorra, Austria, Belgium, Denmark, Finland, France, Germany, Greece, Iceland, Italy, Liechtenstein, Luxembourg, Malta, Monaco, Netherlands, Norway, Portugal, Republic of Ireland, San Marino, Spain, Sweden, Switzerland, and United Kingdom. Geographically, the region contains four climatic zones. These include the circumpolar Arctic in the north, alpine climate in mountainous areas such as the Alps and Pyrenees, a temperate zone in the central area and the Mediterranean zone in the south. Some knowledge of these climatic zones is necessary to understand the ecological factors affecting efforts to promote nature in built space in the region. This is especially true when one considers the critical influence of environmental factors on policy inputs and outcomes. Writing more than a century ago, Edward Hayes (1914) had summarized these factors to include, geographic conditions or natural physical environment, technological conditions or artificial physical environment; psychological conditions, or hereditary and acquired traits of the population; and social conditions or the causal relations between the activities of associates. This chapter is preoccupied with how these and cognate factors affect

A. J. Njoh, *Nature in the Built Environment*,
https://doi.org/10.1007/978-3-030-39759-3_8

efforts to create and maintain nature in built space in Western Europe. For now, suffice to state that, to appreciate the socio-economic capabilities of any area one must first understand the physical conditions that characterize the area. Accordingly, the chapter begins in the next section by examining the geo-ecological, politico-economic, socio-cultural, and historical context of Western Europe. Then, it briefly discusses the religion and its intimate link to capitalism with a view to determining how these affect urban greening initiatives in the region. Subsequent to this, the chapter discusses efforts to create and maintain specific green infrastructure, including forests, farms, energy and water supply in built space throughout the region. The chapter ends with some concluding remarks.

8.2 Politico-Economic, Social, Geo-Ecological and Technological Background

8.2.1 Politico-Economic Factors

Western Europe as discussed in this chapter corresponds approximately with the contemporary European Union (EU). This is distinct from the geographic unit that existed until the creation of the EU in 1993. Prior to 1993, Western Europe excluded Central Europe, which comprised countries of the defunct Soviet Union. Since then, the EU has undertaken a robust initiative to expand into central and eastern Europe. Currently, EU's 28-member-states include erstwhile Soviet Union polities; the complete list includes Austria, Belgium, Bulgaria, Croatia, Cyprus, the Czech Republic, Denmark, Estonia, Finland, France, Germany, Greece, Hungary, Ireland, Latvia, Lithuania, Luxembourg, Malta, the Netherlands, Poland, Portugal, Romania, Slovakia, Slovenia, Spain, Sweden and the United Kingdom. These countries share more than geographical space; they are, to a considerable extent, socio-economically alike. Their economies are generally more developed, while their gross domestic products (GDPs) per capita are typically higher, than the world's average. According to current World Bank data, the European Union is second only to the United States as the world's wealthiest and largest economy (see World Bank 2019). It also contains some of the world's wealthiest economies, including Germany ($4.0 trillion), United Kingdom ($2.8 trillion), France ($2.8 trillion), Italy ($2.0 trillion), and Spain ($1.4 trillion) (Encyclopedia Britannica 2019). Another indicator of EU's prominence as a global economic giant is the fact that it is home to as many as 161 of the world's top 500 largest corporations, that is, the Fortune Global 500. Yet another testament to the economic prosperity of EU member-states is the fact that they currently enjoy, and have historically enjoyed, some of the highest standards of living in the world. However, a few countries in the region such as Greece and some of the erstwhile Soviet polities (e.g. Poland, Romania, Slovenia and the Czech Republic) lag behind their peers economically. Notwithstanding, these countries tend to enjoy the effects of the common social, economic and security policies that govern EU member states. This was essentially an important aim of the Maastricht Treaty of November 1, 1993 that created the European Union in the first place. The Union's common currency, the Euro, seeks to enhance the political and economic integration of member states. Thus, it is reasonable to expect more instances of convergence than divergence in their urban planning policies, particularly those affecting nature in built space. As liberal democracies, EU member-states operate politico-administrative systems that have devolved authority over urban planning and cognate activities to sub-national administrative units. This contrasts sharply with what obtains in authoritarian or other centrally-controlled systems such as polities in which state capitalism prevails. Under such systems, the (national) state is directly involved in urban planning, which it views as a powerful tool to gain access to revenue and exert influence over the society and national economy. In this case, national objectives tend to trump local ones. Yet, the issue of nature in built space like many environmental issues are locally-bound. Thus, and at least in theory, EU member states, because of their propensity to devolve power to local authorities, are better placed to address the problem of nature in built space.

8.2.2 Geo-Ecological Factors

The United Nations Environmental Programme (UNEP) has identified five major environmental issues that are of general concern among EU member-countries. These issues, which are of global magnitude but especially critical in the EU, include (UNEP 2012): air quality, biodiversity, chemicals and waste disposal, climate change, and scarcity of freshwater. As argued in this book, nature in built space possesses the capacity to, among other things, contribute positively to efforts to address these issues. Therefore, it is in order to examine the issues in greater detail.

Air quality problems persist in the EU, where air pollutant concentration levels remain elevated because of the region's high degree of industrialization. European cities contribute 69% of Europe's CO_2 emissions (UNEP 2012). The major drivers of air pollution in the region include, automobile emissions, industrial production plant emissions, including the emissions of nitrogen oxide, volatile organic compounds, anthropogenic sulfur dioxide, and particulate matter. These sources are concentrated in urban centres. Accordingly, the EU pollution problematic is most acute in cities because of the high concentration of people and activities. Yet, cities in Europe, and anywhere else for that matter, are not necessarily bad. In fact, an urban resident consumes significantly less energy than her rural counterpart.

European Union environmental policy makers are patently aware of these problems and have accordingly taken steps to address them. One such step has entailed restrictions on the sulfur content of fuels, and bans on lead in fuels. These have been intended to help reduce air pollution from transportation and related activities. Apart from region-wide policies such as the aforementioned, there have been country-specific initiatives to curb pollution. Sweden's congestion tax, which was enacted in Stockholm in 2007 exemplifies this. The tax essentially penalizes vehicles not running on clean energy entering the city centre during working hours. Finally, there are measures with transnational implications that have been adopted in the region. One example of this is the pan-European Scientific Network of the Convention on Long Range Trans-boundary Air Pollution (CLRTAP). CLRTAP has helped to reduce levels of pollution in the region through its effective monitoring network. Other policies in this regard have included, carbon taxes, emissions trading schemes, promotion of renewable energy systems, and local voluntary initiatives by national and municipal authorities within the European Union. These measures have yielded positive dividends. For example, there has been a substantial decrease in emissions of many air pollutants in past decades; there have also been vast improvements in air quality across the region.

The second major environmental problem affecting Western Europe that can benefit significantly from efforts to promote nature in built space is biodiversity loss. To appreciate the logic inherent in this assertion it is necessary to understand the meaning of biodiversity. It is essentially the richness in variety of a region's stock of animal species, flora and fauna, natural habitats and ecosystems. The rapid disappearance of these essential features of the environment in built space constitutes an enormous challenge in the region. A quarter of the region's species is threatened with extinction while its fish stocks are overexploited. Efforts to promote nature in built space can go a good way in restoring the region's natural ecosystem. Although not specifically focused on built space, there are many pieces of legislation that have been directed towards conserving biodiversity in the region. The EU Strategic Plan for Biodiversity 2011–2020 constitutes a prime example. In addition, EU member-countries have renewed their commitment to: halt biodiversity loss in EU by 2020; protect, value and restore biodiversity and ecosystem services by 2050. These efforts have yielded mixed results. In this regard, EU member states failed to attain their goal of halting biodiversity loss by 2010. On the positive side, net total forest area in the region is rising despite heavy exploitation.

The third environmental issue of urgent concern in the EU is related to waste, particularly chemical waste storage and disposal. This problem is especially severe because of the region's very intense industrial sector, which date back to the industrial revolution. The problem manifests itself in terms of the massive amount of chemical and other wastes that is generated in the region on a daily basis. The problem is magnified in the EU countries that used to be part of the Soviet Union because of the legacy of unregulated industrial waste management system that prevailed before they became part of the EU. However, it would be erroneous to think that the problem is confined to the former Soviet republics. Rather, it is a problem that can be found everywhere in the region albeit in varying degrees. This is so despite aggressive recycling programs that have been adopted across the region. By some accounts, only 55% of the waste, excluding major mineral waste, generated in the region in 2016 was recycled; this was a modest increase over the 2010 figure of 53% (Eurostat 2019: para. 1).

Recognition of the extent to which waste, especially chemical waste, constitutes a threat to life in built space predated the formation of the European Union. In fact, the first mention of the term, 'environment' in the region was in an official European

Commission (EC) document in 1967. This is when the directive on standards for classifying, packaging, and labeling dangerous substances was issued. The First Environmental Action Program was launched in 1973; it established the first framework for EC environmental issues. Currently, 17 multinational agreements exist to address issues of chemical and other waste management in the EU. In addition, the Strategic Approach to International Chemicals Management (SAICM), established in 2006 plays an important role in chemical and other waste management. A multi-stakeholder framework, SAICM has as one of its goals, to ensure the safe management of chemical waste worldwide by 2020.

The fourth major problem whose resolution can benefit from efforts to promote nature in built space in the European Union is climate change. According to the European Environment Agency (EEA 2019, para. 1),

> Climate change is happening now and is expected to continue. Mitigation, by preventing or reducing the emission of greenhouse gases into the atmosphere, can make the impacts of climate change less severe.

The need to combat climate change and its consequences has been widely recognized not only in the European Union but the world at large. In this regard, 195 countries signed the Paris Agreement in 2015. The agreement was essentially, "to keep the increase in global mean surface temperature to well below 2° C, and to limit the increase to 1.5° C." Efforts to promote nature in built space through urban forestry go a good way to lower surface temperatures. Apart from seeking to reduce temperatures across the region, EU authorities have also been pre-occupied with related environmental problems such as greenhouse gas emissions. The long term policy objective in this case has been to curtail these emissions by 90–95% by 2050 compared with 1990 levels. The shorter term objective, which the EU has set is to reduce greenhouse gas emissions levels by at least 40% below 1990 levels by 2030 (EEA 2019).

The EEA forecasts a number of climate-change-related problems, including extreme weather events resulting in hazards such as floods and droughts. These events, which are likely to contribute to the spread of infectious diseases, are projected to occur with more frequency in the future. These problems are not projected to be evenly distributed across the region. Rather, some areas are likely to be more affected than others. The areas most likely to be affected are as follows. The first is Southern Europe and the Mediterranean basin due to a rise in heatwaves and droughts. The second includes the mountainous areas because of the increase in melting of snow and ice. The third includes the coastal zones, deltas and floodplains because of sea level rises, and increasing intense rainfall, floods and storms. The fourth is Europe's far north and the Arctic, which are likely to suffer from increasing temperatures and melting ice. The fifth is the coastal zones, deltas and floodplains, which are likely to be affected by rising sea levels and increasingly intense rainfall, floods and storms.

The fifth environmental problem which can be addressed, at least partially by well thought-out policies to promote nature in urban space, is freshwater scarcity. However, it is important to note that the problem is not solely one of scarcity. Rather, it is partially one of uneven distribution both at the intranational and international distribution levels. This is largely due to differences in precipitation. For example, annual rainfall ranges from 18 inches (45.72 cm) in western Norway to 3–16 inches (7.2–40.64 cm) over much of central Europe and around 4–16 inches (10.16–40.64 cm) in central and southern Spain. Consequently, water must be moved across considerable geographic distances. It is estimated that as much as 40% of the water is often lost before reaching its destination. Other problems relating to water, particularly freshwater, in the European Union are largely functions of the region's population density. The problems can be summarized under the following broad categories, namely overexploitation, excessive use, inability of supply to meet demand, and eutrophication (or contamination from sewage discharge, and contamination from agricultural run-off).

It has never been lost on EU policymakers that their region's water sector faces some significant problems that need to be urgently addressed. Consequently, they have adopted relevant measures to deal with the situation. Some of the earliest efforts in this connection were initiated in 1975, and were mainly concerned with the quality and standards of rivers and lakes that served as sources of drinking water (European Commission 2019a, b). These efforts culminated in the setting of binding quality targets for drinking water as well as the enactment of legislation on fish waters, shellfish waters, bathing waters and groundwaters in 1980. The authorities have also strived to address issues relating to pollution and contamination from urban waste water and agriculture. The task of reviewing extant legislation in this connection was assumed by the Frankfurt Ministerial Seminar on Water in 1988. This was later followed by the second phase of water legislation in 1991; in addition, authorities promulgated the Urban Water Treatment Directive, which made provision for secondary (biological) waste water treatment. In addition, there was the issuance of the Nitrates Directive, which was intended to address water pollution by nitrates from agricultural activities. In 1996, the Directive for Integrated Pollution and Prevention Control (IPPC), which was later renamed, the Industrial Emissions Directive, was adopted. The main aim of this is to address pollution from large industrial installations. A number of noteworthy policy initiatives were undertaken in the 2000s. The first of these was the EU Water Framework Directive, which was adopted in 2000. This required EU member-states to design a system of management for river basins. A shared trait of these policies is their focus on water stressors. Thus, authorities are cognizant of the problems of water stress that confront the region. Here, water stress should be taken to mean, 'pressure on the quantity and quality of water resources' (European Environment Agency 2016, para. 1). This issue often results in serious problems of water shortages, flooding, pollution and ecosystem damage. To be sure, many initiatives have been undertaken to address water stress and related problems in the region. One of these is the first European law on bathing water, which was enacted more than two decades ago. However, where a lot more needs to be done is in the area of policy integration and harmonization. According to the European Environment Agency, the integrated management of water resources constitutes the most promising strategy for addressing EU's water problematic.

Efforts to deal with the environmental problems facing the EU have traditionally focused on the supply as opposed to the demand side of the demand and supply equation. For instance, in the water domain, these efforts have typically sought to improve the effectiveness and efficiency of water sources and resources. However, in the recent past, there has been a switch in approach to demand side management, which essentially focuses on water users, including industries, but particularly people. This has proved to be a more effective water resource management strategy than typical supply-side initiatives. As the European Environment Agency (2016, para. 4) observed, demand-side measures that were implemented in Madrid between 1992 and 1994 reduced water consumption by 25%; this is about the same as a reservoir that provides 100 million cubic meters (100,000,000 m^3) per year. Switching from the use of 9-litre to 6-litre toilet water flushes reduced the quantity of water used for toilet flushing by 10% in the United Kingdom (European Environment Agency 2016, para. 4). As suggested here, demand-side water management strategies strive to modify people's behaviors, especially with respect to their use of natural resources. In the case of water, the aim is to instill in people the habit of reducing losses, using water less wastefully—in short, it entails teaching people to efficiently use water and above all, to be the best stewards of water. Success in this regard depends, to a significant degree, on people's culture, of which religion constitutes a paramount element. Accordingly, it is in order to discuss religion as a prominent feature of European culture.

8.3 Judeo-Christianity, Capitalism and Environmental Stewardship

We have discussed the ways in which Christianity may affect its adherents' relationship with the natural environment in Chapter Seven (on North America). Accordingly, this chapter confines this discussion to the reciprocal link between Christianity and an important feature of Western societies, namely capitalism. One of the most authoritative statements on the line linking Judeo- Christianity to economics and the natural environment is by Max Weber (1864–1920) (see Weber 1958). This renowned German sociologist and political economist was so interested in interrogating this link that he proceeded to pen what had since become a classical piece of work under the caption, *The Protestant Ethic and the Spirit of Capitalism* (originally published in 1905). In this authoritative scholarly piece, Weber sought to explain the link between the successes of Western Protestant Christians and the emergence of capitalism. He advanced the view that Puritan ethics and ideas constituted the foundation upon which capitalism and the institutions necessary to sustain it were built. 'The spirit' in the sense employed by Weber, does not connote a metaphysical sense. Rather, it is intended to capture a set of values, especially hard work that leads to progress. This position later became the basis for the oft-recited maxim, 'Heaven helps those who help themselves.' Although this maxim often passes for a Biblical passage, it is not!

Weber's work focused mainly on Calvinists and other Protestant Christians whom he argued, were instrumental in creating the capitalist spirit. He contended that there is a direct link between adherence to Protestant religion or Protestantism and the propensity to be economically aggressive. For Christians of Protestant leanings, capitalism guaranteed economic success in a manner unavailable in other economic systems. More importantly, capitalism considered profit as an end in its own right; ironically, the spirit of capitalism views the modern pursuit of profit as virtuous. Thus, Protestantism afforded worldly pursuits a religious character; yet, the question of why people as individuals, corporations and governments may elect to indulge in profit-seeking initiatives remains unanswered.

What is the place of nature in all of this? Calvinists and other Protestant sects believe in predestination. In other words, they believe that God has already determined who is saved and who is condemned. In a bid to find clues indicating whether one is already saved or not, Christians of this persuasion developed a profound appreciation for success in worldly pursuits. This desire, and the steps that were initially taken to satisfy it, short-circuited traditional economic systems. This in turn, gave birth to modern capitalism. This, among other things, dictated a need to disregard otherwise cherished Protestant values. Lest this line of argument be misunderstood, it is important to note that no causality is insinuated, let alone asserted, here. Rather, as Weber, contended, Protestantism was simply presented as no more than one of many elements that helped to propel the capitalist system. In the same light, Christianity was also at the root of industrialization and the demise of traditional economies. By traditional economies, we are alluding to the systems typically associated with indigenous societies. Here, people worked just enough to meet their basic needs. In such systems, there was never a need to accumulate wealth and no one aspired for this as it would have been considered greedy. Capitalism is different; it encourages materialism and the glorification of materialism. More importantly, it emphasizes the proverbial 'bottom line,' in other words, the financial profit of any undertaking. This has far-reaching implications for the manner in which nature, and environmental resources writ large are treated in capitalist systems in particular, and the contemporary world in which capitalism dominates in general. According to Lynn White (1967), the contemporary environmental crisis is primarily an upshot of capitalism's dominance. The crisis stems from a misplaced faith in science and technology—the belief that science can help humans achieve any desirable goal. It is, White hastens to remind us, a function of the dominant worldview that places humans and nature in separate compartments and prioritizes humans over nature. It is noteworthy that modern science and technology have their roots in Occidental ethos. This ethos is largely influenced by Judeo-Christian teleology. Ironically, this teleology influenced the emergence of Marxism and Islam. As White (1967) explains it, they are both a Judeo-Christian profanation or heresy.

It is not enough to interrogate avowed Judeo-Christian doctrine to understand how it has influenced environmental thought. Rather, it helps to assess its impact in devaluing the essence of the natural environment in people's lives in non-Western societies. Consider, for instance, the pejorative qualification of indigenous religions by Christian missionaries on the self-proclaimed mission of introducing God as characterized in the Bible to the heathen. In Africa, for example, the missionaries were quick to christen indigenous religions as animism, and sought to annihilate same. They substituted the cult of saints, an action that Africans considered an abomination especially because they considered saints as functionally different from the natural objects they had grown accustomed to revering as sacred. For one thing, saints are people but their citizenship is in heaven. In contrast, the sacred objects that indigenous religions revered are natural and possess spirits of their own. Christian missionaries considered this sacrilege, and moved speedily to discourage such thinking. To the extent that they were successful, natural objects were purged of their spirits, and consequently no longer protected from humans. This, as White (1967: 1205) argues, meant that "man's effective monopoly on spirit in the world was confirmed and the inhibitions to the exploitation of nature crumbled."

However, one would be remiss by not acknowledging the extent to which Judeo-Christianity and commensurate science have promoted knowledge of the natural environment. This knowledge is necessary for conceiving, formulating and implementing sound environmental protection policies. Contemporary science, it is worth recalling, traces its roots to Judeo-Christian ethos. In fact, during its infancy, Western science was almost indistinguishable from Judeo-Christian doctrine. As White (1967: 1206) observed, Western scientists believed their most priced reward was to "think God's thoughts after him." Therefore, with respect to the environment, it is safe to conclude as follows. Environmental science and its subfields, including environmental philosophy, environmental ethics, and environmental policy have evolved from Christian doctrine, particularly Christian attitude toward man's relations to nature. Yet, it cannot be lost on anyone that Christian doctrine views natural objects such as geographical features as simply physical facts. Therefore, notions such as sacred mountains, sacred rivers, *etcetera* that have been instrumental in promoting environmental stewardship in indigenous societies are alien in regions such as those comprising the European Union that are heavily influenced by Christian doctrine. However, it would be erroneous to conclude that such regions do not recognize the importance of nature in built space. As shown in the remainder of this chapter, environmentally friendly initiatives are vibrant in European Union member countries. We focus particularly on efforts to promote nature in built space in the following substantive areas: tree-planting and forestry, food systems, energy, and water generation/supply. The discussion is guided by a set of '5-W'—(What, Why, When, Where, and Who)—questions, as follows:

- What is nature in built space in the context of EU countries?
- Why is nature in urban space important in the EU context?
- When did nature in urban space emerge as an important area of focus in the EU?
- Where did nature in urban space first surface as an issue of concern in the region?
- Who are the major stakeholders with respect to nature in built space in the EU?

8.4 Urban and Peri-Urban Forestry

The entity referred to here as urban and peri-urban forests includes all the trees and woodland in and around urban areas. Such entities are by no means unfamiliar to Europeans. Europe or the 'Old Continent' as it is sometimes called, has a long-standing tradition of urban green space planning and management. The roots of this goes back to the domestication of fire and the development of agriculture writ large. More specifically, these roots are found in the social stratification that occurred during feudalism in Europe. Feudal societies are well-known for concentrating power in the hands of a few namely monarchs and the clergy. A prominent element of the accoutrement of these societal elites of the time were forests where they could retire at will for serenity and tranquility. These forests were not opened to the general public, which was inconsequential because urbanization levels were still considerably low. This meant, among other things that, forests were still relatively abundant. However, over time, urbanization levels in Europe experienced rapid growth and before long, the region had become the most urbanized in the world. Commensurate with the urbanization experience of Europe was the empowerment of cities and their inhabitants. More importantly, urbanization in Europe engendered democracy; this meant that people were more able to express themselves and impress upon the state their wishes and desires. One wish that was frequently expressed was for public green space. In a bid to respond to these wishes authorities in European countries began making provision for trees and green spaces in their urban plans. Thus, as noted below, they essentially followed in the footsteps of North American urban planners who have been at the forefront of the modern movement to promote forestry in built space. Urban forests can be seen as essentially one of the consequences of urbanization. In this regard, many urban forests have evolved over the years as conventional forests that have been incorporated in urban areas through urbanization.

However, it must be noted that for a long time in western Europe, forests have been associated with the country and hardly with urban areas. This was especially the case up until the period of the Industrial Revolution of the eighteenth century. However, by the nineteenth century many parts of Europe were already recognizing forested environs as necessary features of urban areas; these environs, Europeans had come to believe, constitute a key ecosystem service provider (University of Copenhagen 2018). This belief was legitimized and intensified with the acceleration of urbanization in Europe in the twentieth century. Consequently, European countries increasingly took steps to protect woodland landscapes especially because of the multifaceted social and environmental roles they play. Yet, it must be noted that the earliest urban forests in Europe never originated in that capacity. Rather, they originated as traditional forests that were absorbed or fragmented by urban sprawl. Thus, urbanization has had a major impact on European forestry (Konijnendijk 2003). This is especially true with respect to modern urban forestry—that is, the variant of urban forestry, which as mentioned earlier, emerged in the United States, and quickly found its way to Europe. In the 1960s when this renewed concern for urban forests spilled over into Europe, the focus was more on green space in general.

This differs from what obtains in North America, where urban forests have always been taken to include the totality of trees within the urban milieu. In Europe, in contrast, urban forests include mainly forests within or near urban areas. The *Forêt de Saint-Germain* and *Fontainebleau* in Paris, France; the Epping Forest in London, England; and the Grunewald Forest in Berlin, Germany, are good examples of ancient urban forests in Europe.

It is necessary to accentuate the distinguishing mark between urban and conventional forests. Urban forests differ significantly from their conventional counterparts in many ways, but particularly in the way they are, or ought to be, managed (Lawrence et al. Online). In contrast to conventional forest management, urban forest management involves many stakeholders, including state and non-state actors interacting at different levels. Within the framework of the democratic systems of E.U. member countries, entities at all levels of government, including national and municipal governments can impact urban forestry inputs and outcomes. This is essentially because urban forestry activities impact other pieces of infrastructure such as streets, roads, highways and railways. These are usually under the auspices of municipal and national governmental bodies. Also, different interests are impacted because urban forestry activities often require land that is often in individual hands. However, the most influential actors in matters relating to urban forestry in E.U. member countries are municipal authorities. This is especially because of the critical role that urban planning and its ancillary organs play in the design, timing and location of physical objects in space in the region. One feature of urban forests in Europe that distinguishes them from those in other parts of the world is their *raison d'être*; they mainly serve recreational and aesthetic purposes. Here, we hasten to state that urban forests in especially non-Western societies play sacred and sentimental roles. As mentioned earlier, urban forests in Europe are treated more generally as part of the urban green infrastructure. In this regard, countries such as the United Kingdom employ the same policies to govern urban forests and green infrastructure without any distinction (Mell 2010). Also, many of the requirements for protection and participation relate equally to forestry and other aspects of urban green space (Mell 2010). However, as Lawrence and her colleagues (2013) have observed, the literature on urban green infrastructure in Europe seldom makes any allusion to trees. This, they lament, is unfortunate since many of the significant benefits of green infrastructure are derived from trees rather than from other types of vegetation.

Furthermore, there are unique issues circumscribing trees in built space in European countries because of their long history of town planning and its ancillary elements such as zoning and environmental regulation. In this case, there are specific laws, policies and regulations that govern trees in built space in European countries that may be unknown elsewhere. Also, the government institutions responsible for implementing the relevant laws reflect the democratic nature of politico administrative structures of which they are a part. In this regard it is worth noting that the management of urban forestry policy making, like other aspects of urban management, are largely of the participatory variant. In fact, laws governing trees in some European countries date back more than a thousand years (Lawrence et al. Online). The laws originally arose out of a concern over ownership and exploitation rights as well as a concern for matters regarding arbitration about property boundaries and possible dangers presented by trees and woodlands. Yet, it is worth noting that there are some differences among European countries with respect to tree laws. In fact, there are even significant differences among municipalities within certain countries. For instance, in Austria, Vienna boasts the longest history with policies specifically designed to protect trees in modern times. The city's first tree ordinance in modern times date back to 1974.

An important feature of urban forestry worth noting has to do with its avowed purpose. In the United States, for instance, it was proposed as a viable approach to managing natural resources in urban environments. Initially, there was resistance to the concept of urban forestry from many camps, primarily from foresters and urban green professionals in the United States. In Europe, where the purpose was to mitigate the negative externalities of rapid urbanization, urban forestry-promoting efforts also met with similar resistance. It was not until the 1990s that the concept began having broad appeal. This broad acceptance was accompanied by the emergence of a European urban forestry research community. This community has since remained vibrant; and so has the activities to promote urban forestry. Commensurate with this was the introduction of policies, courses, and whole educational programs focusing on urban forestry. Although relatively small in comparison to those of other parts of the world, urban forests in Europe cover millions of hectares of land. They provide multiple, scarce agricultural goods and services.

Another aspect deserving of attention is the implementation phase of urban forestry. This is typically characterized by fragmentation. In Europe, this often means the division of implementation functions among many, sometimes disparate agencies. The day-to-day operational responsibility over tree care is usually left to municipal authorities while the actual planning and enactment of tree laws is placed under the charge of the national government. The degree of power devolution implied here exist exclusively in the traditionally democratic polities of the region. The cases of Denmark, Norway, Finland and Germany, where decision-making powers are in the hands of regional policy makers are illustrative. Conversely, urban planning powers, and by extension, the power to enact laws affecting UPF in the erstwhile communist states in the region remain in the hands of central authorities.

Also worthy of note is the issue of land tenure and its implications for urban forestry. Tenure connotes land ownership rights, claims of entitlement, access and use rights. These rights tend to differ by country, culture and level of politico-economic development. They also differ by political economic ideology such as communist/socialist versus capitalist/democratic polities. Occasionally, there have been differences between polities of the same ideological leanings. For instance, some studies in the U.S. have suggested that urban tree planting activities typically occur on private land (Perkins et al. 2004). In contrast, much of the urban trees and woodlands in Europe are located on public lands. These lands are often owned by municipal governments, and in some cases, by state governments. This is not to say that private ownership of urban trees is non-existent in Europe. Rather, it must be noted that private ownership is prominent when urban forests are defined more broadly to include gardens and green space writ large.

The stakeholder pool that can be found in any urban forestry domain in an E.U. member country is also worth examining. In most countries, the central government is the key player in the urban forestry field. It is usually responsible for conceiving and enacting laws governing trees and woodlands. However, the day-to-day management of these pieces of green infrastructure falls under the charge of local governments. Among the consequences of this arrangement are variations in the standard of management among local governments. In practice this means, among other things, that local governments own and manage urban parks, roadside trees, and woodlands. Given their proximity to the public at large, and urban forest users, municipal authorities are also ideally placed to promote community participation (CP) as a strategy for improving the sustainability of urban forests.

To succeed, the implementation of urban and peri-urban forestry (UPF) must involve individuals with different backgrounds. Thus, urban and peri-urban forestry is interdisciplinary. In Europe, those typically involved in UPF are arboriculturists, landscape architects, urban planners, ecologists, foresters, and community workers. In addition to state, municipal and private entities, non-governmental organizations (NGOs) are also very active in the UPF domain in Europe. These are typically environmental activist organizations, and community groups. Some of these depend on the services of volunteers. Thus, given some of the drawbacks of employing volunteers it follows that their use must be moderated in UPF. Yet, there are highly successful NGOs in the UPF field in Europe.

Forests involving multiple stakeholders make up the most common type of urban forests in Europe. This should be considered a positive development as the importance of enlisting the participation of members of any community in efforts to improve the quality of their physical space can never be overstated. In fact, efforts to development and maintain natural features such as forests in built space are likely to be more successful if they meaningfully involve users of that space. The merits of urban forests, including their utility in contributing to biodiversity conservation, serving as a locale for recreation, improving conditions for groundwater production, mitigating urban heat island effects and improving air quality are considerably well-established. An important next step is to broaden the pool of those who see themselves as stakeholders in efforts to develop and conserve urban forests. Logically, this should include all users of the built environment.

8.5 Agriculture in Built Space

Urban planning in Europe has always pursued two important goals. The first has historically been to promote spatial order entailing the compartmentalization of land use activities in Europe. Accordingly, European urban plans have consistently sought to distinguish between residential spaces and industrial zones; and between urban and rural areas. Within this framework, urban areas have always been designated, and have functioned, as places of business, trade, education, politico-administration and the exercise of power. In contrast, rural areas have served as the locale for agriculture and primary production. The second has always been to promote human convenience. In practice this has meant minimizing distances between activity centres. Thus, reducing the distance between residential space and farmland would be viewed as an important objective of planning. This seemingly 'anti-planning' objective of planning is best illustrated by Ebenezer Howard's Garden City model (Lohrberg et al. 2016). In its classical form the model was intended to marry the best attributes of a rural environment with those of an urban milieu, thereby avoiding the pitfalls of both. Guided by this important tenet, Howard (1898) proceeded to propose an ideal garden city as one containing no more than 32,000 persons, located on a piece of land with an area of 6000 acres (i.e., 2400 ha.). To assure the residents' convenience, he proposed a concentric spatial organization pattern. The 2400-hectare site was supposed to be self-contained to include open spaces, parks, and wide streets. Most importantly for the purpose of this discussion, the Garden City as the name implies, was supposed to contain farmland, residences, and industries, all of which must be circumscribed by greenbelts.

Howard's model appears to have been a logical, if only utopian, response to the disconnection of cities from rural areas occasioned by the Industrial Revolution of the nineteenth century. In fact, Howard's plan is arguably the best-known attempt to reconnect the rural to the urban. Doing so today is tantamount to undoing a grave error—that related to disentangling urban from rural areas—that was committed in the early twentieth century. The number of Garden Cities that have been developed in many parts of the world constitutes a testament to the model's popularity.

Although seemingly new, the current interest in urban and peri-urban agriculture in Europe is by no means novel. If anything, it can best be characterized as akin to the proverbial 'old wine in new bottles.' This is especially so when one considers the classical works of Heinrich von Thunen (1826) in agricultural economics. Reverberations of von Thunen's pronouncements are still traceable in contemporary debates on land use and urban development, sustainable land use, economic competitiveness, quality of life, food security, and sovereignty, adaptation to climate change and resilience (AGRI Committee 2018). The main driver of urban and peri-urban agriculture, namely urbanization has remained unchanged since the Industrial Revolution. Yet, there are some peculiarities regarding urban and peri-urban agriculture in Europe today. Here, the central objective of urban and peri-urban agriculture are better classified as environmental, social and aesthetic than economic and survival strategies. In addition, European Union states are also interested in promoting urban agriculture as a means of preserving biodiversity, tackling the problem of food waste, and reducing energy consumption. A distinguishing mark of the specific activities that Europeans have summoned to promote and foster urban agriculture is their technological sophistication.

The same technological innovations that have been occurring in communications, health and industrial engineering are beginning to make their presence felt in the agriculture sector. This development has many positive implications for urban farming especially because of its ability to produce much food on very little or no land at all. The definition of urban agriculture proffered by Silicon Canal, a leading

English language technology media source based in Europe captures this strength of urban agriculture most succinctly. According to Silicon Canal, "urban farming is a concept of modern agriculture that involves the adoption of various emerging technology trends including robotics to make agriculture more productive, profitable, and sustainable. The advancement makes it possible for high quality food to be produced at minimal cost. A number of startups in Europe have already proved the feasibility and practical workings of this innovations (Silicon Canals 2019). For illustrative purposes, we present some examples from each of the European countries that have registered the most progress in this regard. These include, France, Germany, Finland, UK, and the Netherlands.

France At least three of the leading hi-tech agricultural establishments currently involved in urban farming are located in France. These include Ynsect, Agricool, and Naio Technologies. Ynsect is an agri-tech that was founded in 2011 to breed insects to produce premium ingredients for organic plant fertilizers, pet food and fish feed. It is the largest insect producer in the world. It is on track to produce about 20,000 tons of protein annually. Agricool is another agri-tech enterprise. It is the inventor of a novel sustainable food growing technique; its main product is strawberries, which it grows on small urban farms using old shipping containers. Known as 'cooltainers,' the containers use LED lights powered completely by renewable energy sources. Apart from using renewable energy, which is less expensive and more sustainable than energy from conventional sources, this technique requires very little water to function. Thus, the technique is far more cost-effective and more sustainable than conventional farming. Naio Technologies is yet another French-based agri-tech enterprise. It came into existence in 2011, and produces and markets agricultural power tools and robots, which are designed to help farmers grow and harvest food.

Germany This country is home to at least two of the world's leading agri-tech enterprises. These include, Infarm, and Agrilution. Based in Berlin, Infarm operates more than fifty modular farms that are designed to be flexible enough to fit just about anywhere. Typically, the units are arranged tightly to be placed together in a confined space. As for Agrilution, it is an agri-tech enterprise that drew its inspiration from the concept of vertical farming. Its avowed aim since its inception in 2013 has been to use the least possible amount of space to grow and market a lot of fresh foods.

Finland One of the leading agri-tech enterprises in the world is located in Finland. Known as Plantui, the enterprise specializes in the development of smart gardens. Its main focus is on indoor gardening, a growing global trend. Its main products to date include smart hydroponic gardens. These are user-friendly as they permit any willing party to grow clean, palatable and healthy vegetables.

United Kingdom The UK is home to Lett Us Grow, an agri-tech enterprise that was established in 2015. It employs aeroponic technology to grow shrubs and herbs. The aeroponic technique, which it uses, boasts a growth rate that exceeds traditional techniques by 70% (Silicon Canal 2019: Para. 3). The technique suspends plant roots in a nutrient-rich mist as opposed to soil. This ensures faster and more consistent results. In doing so, the technique reduces water usage by 95% in comparison to traditional methods. In addition the need for chemicals and pesticides is eliminated thereby minimizing the carbon footprint of urban agriculture.

The Netherlands Holland is the home of The New Farm, an agri-tech enterprise that was established in 2017. It is located in a business building in the Hague. Its avowed aim is to foster and promote vertical farming. It has been set up in such a way that it frees much space on the ground floor for catering and retail (Box 8.1).

Box 8.1 The Roots of Modern Urban Agriculture in Europe

Urban agriculture, also known as urban farming or urban gardening, has a long history in the world in general and Europe in particular. By some accounts, the practice goes back more than four millennia (Viennashares 2019). However, in the case of Europe, it is often associated with the period immediately subsequent to the Industrial Revolution. Here, it is worth noting that prior to this period, and particularly before the advent of railways and steam ships in the nineteenth century, it was not uncommon for large human settlements to designate certain areas within these settlements as farmland. It was here that the settlements cultivated food to meet the basic food needs of their populations. Over time, and with increasing urbanization levels and the concomitant heightened demand for land, it became difficult to set aside any parcel of urban land for anything other than building and cognate use. Consequently, people were compelled to conduct farming and commensurate activities in areas far removed from their settlements. This essentially marked the onset of people's physical separation from their food sources. The 1800s ushered in a new era; it was a high point in the industrial age and urban planners were preoccupied with efforts to protect the health of inhabitants of increasingly densely populated cities. They were also preoccupied with combatting the negative externalities of industrialization. The segregation or compartmentalization of land use activities under the rubric of modernist urban planning was in vogue. Zoning and cognate land use regulation tools were deployed to ensure this land use pattern. However, by the second decade of the 1800s, and with the emphasis on sustainability of large cities, Ebenezer Howard (1898) proposed the garden city as a challenge to the compartmentalized structure of urban centres. This led to the birth of the garden city, a model of urban planning in which gardens and other facilities necessary to promote healthy living co-existed harmoniously. The model soon won the hearts and

(continued)

Box 8.1 (continued)

minds of planners and municipal authorities throughout Europe and the rest of the world. In Europe, much emphasis was placed on the 'garden element' of garden cities especially because of the rapid loss of farmland occasioned by the speedy rate of urbanization that the region was experiencing. Carefully planned urban gardens were summoned to mitigate this loss. In Paris, efforts in this connection in the nineteenth century were simply a continuation of earlier urban planning schemes. These efforts were developed on the foundation that had earlier been established by the Marains system. The Marains, that is, the Swamp system, was so-called because it involved literally farming in swamplands. Such farming had been taking place in France from time immemorial. The crops are usually grown beneath a glass roof and protected with straw from harsh weather conditions (Viennashares 2019). In Germany, efforts to promote urban gardens took place under the canopy of allotment gardens (or *Schrebergarten*) since the 1860s. These efforts were later to find their way into Austria in the twentieth century. However, it was not until the outbreak of the world wars that the efforts recorded significant gains. This was in response to the acute need for food that was caused by the wars. Similar gains were recorded in other parts of Europe where Victory Gardens were established to increase food supply. Urban gardens also found favor with Viennese settlers displaced during the inter-war period. Viennese settlers are known to have reacted to the scarcity of housing and food after World War II by building their own residence and gardens which targeted autarky and sustainability. Initially, this was illegally carried out but later they gained the approval of the magistrate. Allotment gardens, as these were called, continued to constitute a conspicuous feature of the urban landscape in some parts of Vienna.

Source: Viennashares.org.

8.6 Energy in Built Space

The European Union (EU) depends on several sources for the energy it consumes. These include some EU-member states, and foreign countries. Foreign sources account for as much as 55% of the energy consumed within the Union, while only 45% of the energy is domestically produced (European Commission 2019a). The energy mix within the Union is worthy of note. In 2017, the range of energy sources included crude oil, which contributed 36%; natural gas, which made up 23%; solid fossil fuels, comprised 15%; renewable energy, which made up 14% of the total energy consumed, and nuclear energy, which contributed 12% to the energy mix. This suggests that the EU continues to largely depend on conventional energy sources, and renewable energy accounts for only 14% of the energy supply. However, it is important to note that EU member countries differ considerably in terms of their energy mix and energy supply profiles. For instance, petroleum products including crude oil constitutes the source of most of the energy consumed in some EU countries. This proportion is as high as 93% in Cyprus, 88% in Malta, and 64% in Luxembourg. Natural gas, another important source of energy in the region, makes up a third of the energy consumed in Italy, the Netherlands, and the United Kingdom. Solid fossil fuels make up 71% of Estonia's energy, 48% of Poland's, 40% France's, and 31% of Sweden's. Latvia and Sweden get as much as 40% of their energy from renewable energy sources.

The European Commission is at the forefront of energy policy making initiatives in the European Union. In February 2015, the Commission launched a new strategy for a resilient energy Union. The Union is charged with the responsibility of giving EU energy consumers, mainly households and businesses, sustainable, competitive and affordable energy (European Commission 2019a). As a strategy, the Union comprises five inextricably intertwined and mutually reinforcing dimensions whose avowed aims are to propel greater energy security sustainability and competitiveness. The dimensions include (European Commission 2019a): energy security, solidarity and trust, which calls for diversifying the region's energy sources and improving the performance of locally produced energy; a fully-integrated internal energy market, which is necessary to facilitate the free flow of energy among EU-member states; energy efficiency with a view to moderating demand, which is important in efforts to reduce pollution from energy consumption; decarbonizing the economy, which is an element in the push towards a global deal for climate change and efforts to encourage private investment in new energy technologies; and research, innovation and competitiveness, which is necessary to support breakthroughs in low-carbon technologies through research and development in partnership with the private sector.

8.7 Water in Built Space

Europeans have a very high demand for water. The region uses billions of cubic metres of water annually for domestic, agricultural, manufacturing, and industrial activities. This water comes from many sources, including freshwater lakes, rivers and underground. While at first sight these sources appear to contain a limitless quantity of water, it must be noted that all natural water sources in the region are under threat from many factors. Prominent among these are, population growth, urbanization, pollution and climate change-related factors such as droughts, floods, and forest fires. Some have opined that Europe may be on track to face the level of water stress, that is, excessive pressure on water quantity and quality, that is menacing cities such as Cape Town, South Africa and Cairo, Egypt. In fact, water stress already affects as many as one hundred million people in Europe today (EDJN 2019). The region's ground water sources are already under threat from overexploitation, pollution, and climate change.

The problems confronting water supply initiatives in Europe are not different from those facing any other resource situation in which demand exceeds supply. In the specific case of water supply in Europe, the problem is a function of two specific factors. The first has to do with water quantity, which is a function of overexploitation or droughts. The second relates to deteriorating quality, which is caused by pollution and eutrophication. One point worth accentuating with respect to water in Europe is the uneven distribution of available economically accessible water resources. Resulting from this has been major difference in levels of water stress across sub-regions and seasons. Apart from natural forces, there are man-made factors at the root of Europe's water problematic. At the top of the list of these is population growth and commensurate urbanization. This has significantly increased the demand for water over the years. Without an increase in water supply sources, this has meant, *inter alia,* the dwindling of renewable water resources per capita. In fact, E.U. countries such as Portugal, Spain, and Greece have already been experiencing historically high incidences of droughts in recent years. According to some sources, renewable water resources have declined by 24% since the last couple of years. Such a decline has been recorded mostly in the southern European countries. This is a function of decreased precipitation throughout the southern sub-region. The European Environment Agency (EEA) registered a decline of 20% in forest water resources, including lakes, rivers, groundwater, and other reservoir between 2014 and 2015 (EEA Online).

Water scarcity problems have already hit parts of northern Europe, which have historically been spared the brunt of these problems. In particular, countries such as the United Kingdom and Germany have been experiencing serious droughts during the summer months in the recent past. Water stress and related problems are also threatening agriculture-dependent European sub-regions that require irrigation to be productive. Similarly, tourist attractions are witnessing signs of water stress. These problems are on track to be intensified by climate change effects. Therefore, meaningful efforts are called for to mitigate the situation. Some of the steps that have been taken in this regard include efforts to improve efficiency in water use pursuing better water management strategies. The success of initiatives in this regard hinges tightly on the ability to incorporate built space as a critical element in environmental management. Within the politico-administrative framework of the E.U., member states are encouraged to develop and implement better water management policies. One of the aspects of water management that have attracted the attention of E.U. authorities is water pricing policy. In addition, the E.U. authorities have also advocated the adoption of policies to promote public awareness with a bid to improve efficiency in the use of water.

As suggested here, water constitutes one of Europe's major environmental problems. However, it is necessary to reiterate the fact that the problem is not equitably distributed. This is a reflection of the uneven distribution of renewable freshwater sources across the region. On the whole, European countries boast a lower level of water availability in comparison to other regions of the world. One-third of these countries has less than 5000 cubic metres of water per head per year (EDJN 2019). Some countries in the region fall significantly below this mark. For instance, Malta has only 100 cubic metres of water per head per year.

Throughout modern history, Europe has been known for its economic and industrial might. It is essentially for this reason that water takes on a different level of importance in the region. Water constitutes a crucial input in Europe's industry and economy in general. Europe's agriculture sector alone uses up to 40% of all water used in the region per year. This is the most quantity used by any single sector in the European economy as a whole. Thus, agriculture constitutes, and is projected to maintain its position as, the highest consumer of water in the region (EDJN 2019). By implication, the agriculture sector is leading cause of water stress in the region. It is also the sector likely to suffer the most from any water shortage problem in the region. This is due to the sector's excessive dependence on irrigation.

One important, but oft-ignored water-dependent sector of the European economy is energy. This sector is especially important for the purpose of the present chapter because of its relevance for the built environment. An increasing quantity of water is required to produce energy of both the conventional and renewable variety. This sector alone accounts for at least 28% of the used yearly in the region. This water goes into the cooling of nuclear and fossil fuel-based power plants. Some of the water is necessary to provide hydro-electric energy. Yet another water-dependent sector is mining. This sector accounts for as much as 18% of the annual water used in the region. The water usage most relevant for the purpose of this chapter is the household. This usage accounts for about 12% of the water used in the region per year. The daily water supply to households in the region is 144 liters per person per day (EDJN 2019). Although often ignored in the discourse on water use, households also tend to consume water indirectly. This is usually through their consumption of energy such as electricity, which typically requires large quantities of water to produce. This essentially characterizes what the International Energy Agency (IEA) has described as the 'independence of two critical resources' (IEA 2019). An understanding of this interdependence is necessary to fully appreciate the importance of water in the built environment of highly urbanized locale such as Western Europe.

Water Supply and Sanitation Policies in Europe Until the twenty-first century, water supply policies in the E.U. fell mainly under the charge of each member state. Since the onset of this century, a number of region-wide policy actions have been taken. At least seven of these deserve more than passing mention here. The first is the 1975 Directive on surface water quality for drinking water absorption (75/440/EEC). The second is the Bathing Water Directive (BWD/76/160/EEC). The third is the Ground Water Directive of 1979 (GWD/80/68). The fourth is the Drinking Water Directive (DWD80/778/EEC). The fifth is the Nitrate Directive (NiD, 91/676/EEC). The sixth is the Urban Waste Water Treatment Directive (UWWTD, 91/271/EEC). The seventh is the Water Framework Directive of 2000 (DFD, 2000/60/EEC). This directive is distinct, and stands out from the others because it is the first to introduce a holistic approach to ecosystem-based management. It is contemporaneously concerned with the multiple causes of pollution, the modification of flow regimes through abstraction or regulation and morphological alterations and any other form of exposure.

8.8 Concluding Remarks

Many political, economic, social, technological, ecological, cultural and historical (PESTECH) factors affect efforts to promote nature in built space in western Europe, which is considered in this chapter to approximately correspond with the European Union. The politico-economic and socio-cultural context of the European Union has far-reaching implications for efforts to promote nature in built space in the region. Here, it is important to note that the EU has a strong central government structure, which by no means usurps the powers of EU-member states and their respective municipal authorities. Within this framework, the central government is well positioned to deal with trans-boundary environmental issues that affect the region as a whole. Examples of these include energy, water and pollution. In contrast, EU member states possess the power to handle environmental matters of national and municipal scales. Examples include matters involving hygiene, sanitation, and the generation and distribution of potable water.

The ecological and related factors with implications for nature in built space manifest themselves as either problems or solutions; in some cases, they articulate themselves as both. Consider the ecological and technological factors, for example. Issues of air quality, biodiversity, the disposal of chemicals and waste as well as the scarcity of freshwater, are a function of nature or ecology and level of technological development. On its own, technological innovation is incapable of addressing problems such as freshwater scarcity, air pollution and biodiversity loss. Rather, natural measures such as the greening of built space are necessary to complement technology. This line of thinking is most apropos in the European context because of the region's long history with industrialization. As noted above, air pollution remains one of the region's most nagging problems. This is most acute in urban areas, thereby lending credence to any effort designed to promote nature in built space as called for in this chapter. Here, we hasten to note that the leading sources of pollution and environmental contamination in the EU, including automobiles and industries, are located in built space. The environmental problem most likely to be directly addressed by initiatives to promote nature in built space is biodiversity loss. This is because aggressive efforts can succeed to restore the region's stock of animal species, flora and fauna, natural habitats and ecosystems. As noted in this chapter, these resources are threatened with extinction. Another set of actions necessary to complement technological innovation is through well-thought-out policies; some of these have been enacted by EU policy makers. One example that comes to mind is by placing restrictions on the sulfur content of fuels and bans on lead in fuels. Another entails taxing environmentally adverse behavior. Sweden's congestion tax, which penalizes vehicles not running on clean energy entering the city centre during working hours on weekdays, constitutes a good example. Other examples include carbon taxes, emissions trading schemes, promotion of renewable energy systems, and local voluntary initiatives by national and municipal authorities within EU-member states.

References

AGRI Committee (2018) Agriculture and rural development—research for AGRI committee urban and peri-urban agriculture in the EU. Policy Department for structural and Cohesion Policies. Directorate General for International Policies of the Union. PE 617.468, April 2018

EEA (Online) Mean precipitation. European Environmental Agency (EEA) (2019). Available Online. Accessed 3 Nov 2019 via: https://www.eea.europa.eu/dataand-maps/indicators/european-precipitation-2/assessment

EDJN (2019) Water stress in Europe (EEA). European Data Journalism Network (EDJN). Online publication accessed 5 Nov 2019 via: https://www.europeandatajournalism.eu/eng/News/Useful-data/Water-stress-in-Europe-EEA

Encyclopedia Britannica (2019) European Union: Union Organization. Accessed 23 Oct 2019 via: https://www.britannica.com/topic/European-Union

European Commission (2019a) Shedding light on energy in the EU: a guided tour of energy statistics (2019 Edition). Accessed 3 Nov 2019 via: eu.europa.eu

European Commission (2019b) Environment: introduction to the EU water framework directive. Available Online. Accessed 23 Oct 2019 via: https://ec.europa.eu/environment/water/water-framework/info/intro_en.htm

European Environment Agency (2019) Climate change mitigation. Accessed 23 Oct 2019 via: https://www.eea.europa.eu/themes/climate

European Environmental Agency (2016) The problem of water stress. Available online. Accessed 23 Oct 2019 via: https://www.eea.europa.eu/publications/92-9167-025-1/page003.html

Eurostat (2019) Circular economy in the EU record recycling rates and use of recycled materials in the EU, Eurostat Newsrelease, 39/2019–4 March. Accessed 23 Oct 2019 via: https://ec.europa.eu/eurostat/documents/2995521/9629294/8-04032019-BP-EN.pdf/295c2302-4ed1-45b9-af86-96d1bbb7acb1

Hayes EC (1914) Effects of geographic conditions upon social realities. Am J Sociol 19:813–824

Howard E (1898) Tomorrow: a peaceful path to real reform. Swan Sonnenschein & Co., London

IEA (2019) Energy and water: exploring the interdependence of two critical resources. Available online. Accessed 3 Nov 2019 via https://www.iea.org/topics/energy-and-water

Konijnendijk CC (2003) A decade of urban forestry in Europe. Forest Policy Econ 5:173–136

Lawrence A, Johnston M, Konijnendijk CC, Vreese RD (Online) The governance of (Peri-)urban forestry in Europe: briefing paper 3. Accessed 25 Oct 2019 via: https://ec.europa.eu/agriculture/sites/agriculture/files/events/2011/urban-peri-urban-forestry/lawrence_en.pdf

Lawrence A, De Vreese R, Johnson M, Sanesi G (2013) Urban forest governance: towards a framework for comparing approaches. Urban For Urban Green 12:464–473. https://doi.org/10.1016/j.ufug.2013.05.002

Lohrberg F, Licka L, Scazzosi L, Timpe A (2016) Urban agriculture Europe: agriculture interacting with the urban sphere. Jovis

Mell (2010). Green infrastructure: concepts, perceptions and its uses in spatial planning. Thesis, School of Architecture, Planning and Landscape, Newcastle University, Newcastle Upon Tyne, UK

Perkins HA, Heynen N, Wilson J (2004) Inequity in an urban reforestation program: the impact of housing tenure on urban forests. Cities 21(4):291–299

Silicon Canal (2019) Making farming smarter: these 10 European agtech start-ups use robots to grow food! Silicon Canal News, May 22, 2019. Available online. Accessed 5 Nov 2019 via: https://siliconcanals.com/news/making-farming-smarter-these-10-european-agtech-startups-use-robots-to-grow-food/

UNEP (2012) Global outlook on sustainable consumption and production policies taking action together. United Nations Environmental Programme (UNEP). Available Online. Accessed 4 Nov 2019 via: https://sustainabledevelopment.un.org/content/documents/559Global%20Outlook%20on%20SCP%20Policies_full_final.pdf

University of Copenhagen (2018) Urban forests in a European perspective: what can national forest inventory tell us? Workshop for practitioners and researchers held in Brussels: workshop results, April 2018

Viennashares (2019) History time! Urban agriculture—let's look at some roots. Accessed 29 Oct 2019 via: viennashares.org

Weber M (1958) (originally published, 1905). Protestant ethic and the Spirit of capitalism. Scribner, New York

White L (1967) The historical roots of our ecological crisis. Science 155:1203–1207

World Bank (2019) GDP (Current US$) European Union, United States, China. World Bank data, available online. Accessed 29 Jan 2020 via: https://data.worldbank.org/indicator/NY.GDP.MKTP.CD?locations=EU-US-CN

The manufacturer's authorised representative in the EU is Springer Nature Customer Service Centre GmbH, Europaplatz 3, 69115 Heidelberg, Germany. If you have any concerns regarding our products, please contact ProductSafety@springernature.com

Printed and bound by CPI Group (UK) Ltd, Croydon, CR0 4YY
30/09/2026
02225453-0012